THE RIDE TO MODERNITY

The Bicycle in Canada, 1869–1900

This is the story of Canada's encounter with the bicycle in the late nineteenth century, set in the context of the cultural movement known as 'modernity.' Glen Norcliffe covers the bicycle's history from about 1869, when the first bicycle appeared in Canada, until about 1900, a date that marks the end of the era when bicycles were a recognized symbol of modernity and social status. Cycling continued into the Edwardian period and beyond, of course, especially in Europe, but by then it had lost its symbolic status and social cachet in Canada

Norcliffe's aim is to examine how the bicycle fits into the larger picture of change and progress in a period of dramatic economic, social, and technological flux. He argues that the bicycle led to a host of innovations affecting the development of technology, modern manufacturing, better roads, automobiles, and even airplanes, He describes, for example, how the bicycle, promoted through eye-catching advertisements, was one of the first products for which the sale of accessories was as important as the sale of the main item – thus anticipating twentieth-century patterns of marketing.

Lively and well illustrated, *The Ride to Modernity* provides a particularly Canadian history of one of the first big-ticket, mass-produced consumer luxuries

GLEN NORCLIFFE is a professor of geography at York University

The Ride to Modernity

The Bicycle in Canada, 1869–1900

Glen Norcliffe

UNIVERSITY OF TORONTO PRESS
Toronto Buffalo London

Toronto Buffalo London

Printed in Canada

ISBN 0-8020-4398-4 (cloth)
ISBN 0-8020-8205-X (paper)

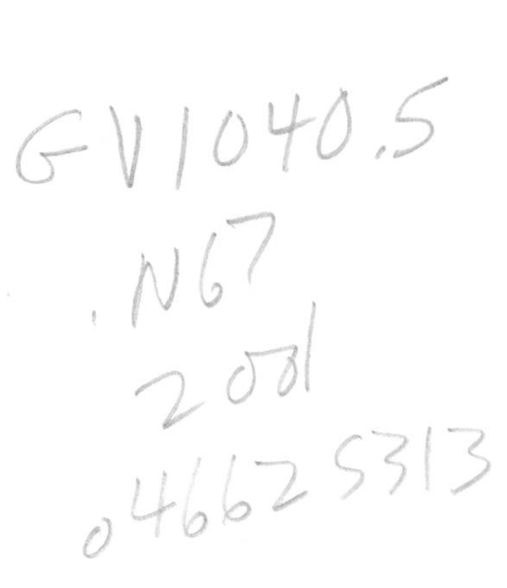

Printed on acid-free paper

National Library of Canada Cataloguing in Publication Data

Norcliffe, Glen
The ride to modernity : the bicycle in Canada, 1869–1900

Includes bibliographical references and index.
ISBN 0-8020-4398-4 (bound) ISBN 0-8020-8205-X (pbk.)

1. Bicycles – Canada – History – 19th century. 2. Cycling – Canada – History – 19th century. 3. Cycling – Social aspects – Canada. 4. Bicycle industry – Canada – History – 19th century. I. Title

GV1040.5.N67 2001 796.6'0971 C2001-930398-4

University of Toronto Press acknowledges the financial assistance to its publishing program of the Canada Council for the Arts and the Ontario Arts Council.

This book has been published with the help of a grant from the Humanities and Social Sciences Federation of Canada, using funds provided by the Social Sciences and Humanities Research Council of Canada.

University of Toronto Press acknowledges the financial support for its publishing activities of the Government of Canada through the Book Publishing Industry Development Program (BPIDP).

For Mary

Contents

Illustrations

Preface

To use a bicycling metaphor, this book spins on two wheels, the one being academic and the other social. The front wheel, which is the academic one, has guided me down the path I have taken in writing the book. My research as an industrial geographer has been focused for many years on the changing location of industry, on forms of work, on globalization, on trade and consumption. The theme of modernity, around which the book is organized, connects to all of these interests and, of its very nature, introduces cultural issues into the economic analysis. I hope to persuade the reader that there are valuable lessons to be learned from viewing industrialization as part of the complex narrative of modernity.

The rear wheel of my metaphorical bicycle, the personal one, is the driving wheel that kept the project rolling. As a cyclist who rides machines both ancient and modern, I have had the pleasure on many occasions of accompanying cycling enthusiasts who have shared their extensive knowledge of bicycles and tricycles with me. These conversations and my reading into cycling history led me to conclude that there was a bigger story to be told about the evolution of bicycles than the usual Whig history of technological progress.

To pursue the metaphor, I began to see that a frame could be constructed connecting these two wheels. It was one of those rare occasions, and the first in my own research experience, where a personal interest and my academic work converged. I realized that many of the concepts deployed in my research and teaching in contemporary industrial geography had direct relevance to bicycling history, with the potential to provide a more discursive history of bicycling. When I started to read the papers presented at the annual International Cycling History Conferences inaugurated by Nick Clayton in 1990, I discovered that there were several other researchers who shared a similar viewpoint. Nicholas Oddy, in particular, has made the case that bicycling history needs to be situated in larger social, economic, political, cultural, and geographical debates.

In chapter 1 of this book I introduce the concept of modernity, which informs the whole discussion. It is a theme that I return to on numerous occasions in subsequent chapters and that I re-examine in the final chapter. The wave of interest in cycling is interpreted as an episode in the project to modernize daily life. One of the principal mechanisms by which modernity spreads is the carrier wave, which is discussed in chapter 2. Within the carrier wave concept several components that advance modernity are described. The argument here draws on an analysis of Canadian bicycle patents to illustrate the flow and ebb of interest in bicycle-related inventions. In chapter 3 I turn to methods of bicycle production by examining the rise of industrial modernity in the workshops and factories that produced Canadian bicycles. Since production and consumption are seen as but two sides of the same coin, chapter 3 is mirrored in chapter 4 by an examination of how Canadians 'consumed' bicycles.

The modernizing impacts of the carrier wave associated with the bicycle extended into several related spheres of life. The connections with road improvement are considered in chapter 5. This is an important theme, since many carrier waves have required investment in social capital for the process to take off. In chapter 6 the social impacts of the bicycle and its connection to social modernity are assessed. Collectively, these changes produced new geographies associated with the enlarged spaces and particular places where bicyclists made themselves visible. The notion of the cycling flâneur is suggested in the conclusion to chapter 7 as a quintessential expression of modernity. In the final chapter, entitled 'Pedaller's Progress,' the discussion turns to the question of what the bicycle boom adds to our understanding of modernity.

I recognize the danger of trying to write for two audiences and pleasing neither. This essay is addressed to an academic audience, but my hope is that the contents will also be of interest to, and accessible to bicycle enthusiasts. The academy should connect with a larger interested public. I also want to stress that information on cycling in Canada in the nineteenth century is still fragmented and dispersed, with many serious gaps in the record; I invite readers with knowledge of early published material, bicycles, and related artefacts to contact me at the Department of Geography, York University, Toronto, Ontario M3J 1P3, so that the work of compiling a more complete record of Canadian cycling can be advanced.

I acknowledge with thanks the Humanities and Social Sciences Federation's programme of support for scholarly publications for its contribution towards the cost of publishing this manuscript. I also include in this acknowledgment my thanks to the two anonymous reviewers of HSSF, whose incisive comments on the manuscript have improved it in numerous ways. The advice and support of Gerald Hallowell and Jill McConkey of the University of Toronto Press in steering the manuscript through many hurdles is gratefully acknowledged. Catherine Frost's careful copyediting removed numerous errors. Kathy Armstrong at York University on numerous occasions, and with great patience, has straightened out my frenetic attempts to do things with text files that Wordperfect does not allow. John Dawson has patiently applied his superb photographic skills to create publishable copies of the illustrations.

The social side of this project kept it going. Above all, Ron Miller has given the entire enterprise immense assistance. He has freely made available his own material on Canadian cycling and, more important, has patiently explained many of the technicalities of bicycles and cycling to the most impervious of students. We have talked over many of the themes on rides in places as diverse as the west of Ireland, the Ozarks of Missouri, and Pelee Island. Lorne Shields made available illustrations from his superb collection of Canadian bicycling ephemera. Jim Cameron read through the manuscript, and covered it with several hundred post-it notes, all of which led to improvements. Nick Clayton, Nicholas Oddy, and Derek Roberts, all three immensely knowledgeable cycle historians, read the first draft of the manuscript and suggested many corrections and improvements (I hope, as a result, that Derek's correction sheet will be considerably shortened)! Most important, my wife, Mary, has also come along for the ride. She has happily joined in the social side of riding, she has tolerated the many evenings and weekends when writing took me out of circulation, and she has commented on drafts of the book.

All of the following also have contributed, in too many ways to list, to this essay: Don Adams, Richard Anderson, Carl and Clarice Burgwardt, Gladys Fong, Cyril Fry, Bill Gladding, Conrad Heidenreich, Bill Humber, Dr Luke Irwin, Henry, Marion, and George Kalbfleisch, Carol King, Jim Lemon, Peter Matthews, Charles Meinert, Hilary Norcliffe, Brian Osborne, John Radford, Andrew Ritchie, Ron and Betty Pequegnat, Carolyn Randall, Jim Spillane, Hap Steinham, Glyn Stockdale, Larry Strung, Marvin Thomas, John Warkentin, J. David Wood.

ARCHIVES: Canadian National Archives: Joan Schwartz, Peter Robinson, Dianne Martineau, Myron Momryk; McCord Museum (Montreal) - Notman Collection: Nora Hague; City of Vancouver Archives: Janette MacDougall; Town of Collingwood Museum: Tracy Marsh; Prince Edward Island Provincial Archives: Marilyn Bell; Dawson City Museum: Sharon Bergey; Yukon Archives: Heather Jones; Stratford-Perth Archives: Carolynn Bart-Riedstra and Brandi Borman; National Museum of Science and Technology: David Monaghan; Provincial Archives of New Brunswick: Fred Farrell; New Brunswick Museum: Regina Mantin; National Map Library: Ed Dahl.

GLEN NORCLIFFE
Maple, Ontario

THE RIDE TO MODERNITY

The Bicycle in Canada, 1869–1900

CHAPTER ONE

Modernity and the Bicycle

It was probably in the late summer of 1869 that coal miners on the day shift at the Caledonia mine in Glace Bay, Nova Scotia, stumbled out into the bright light of day to see the strangest of scenes. A young man – Henry S. Poole, the mine manager's son – was riding around in the mill yard on a curious, two-wheeled, pedalling machine. This was, quite possibly, the first bicycle ridden in the province that, only two years earlier, had been one of the founding members of the Canadian Confederation. Henry's father had taken a holiday in England during that summer, where he had been impressed by a new sensation on the streets, the boneshaker bicycle or, as it was then called, the velocipede. It would seem that he could not resist buying one to take home to his teenage son.[1]

At about the same time in Perth, Ontario, another young Canadian – John Kerr – had been looking for a job. With the help of his father he had found employment with a local photographer. While his new employer was absent one day, Kerr began to experiment with the photographic equipment and took a self-portrait that may be the earliest surviving photograph of a bicycle in Canada (figure 1.1). We do not know who owned the boneshaker, but we do know how Kerr set up the scene in his employer's studio, and we also know how he took the photograph.[2] He positioned the boneshaker against the balustrade, focused the camera on it, and, after draping the camera with a thick black cloth, raised the

Figure 1.1. John Kerr mounted on a velocipede in a photographer's studio at Perth, Ontario, probably in 1869.

exposure cover. Next, he mounted the boneshaker and when he was comfortably seated, he jerked on a length of string, which pulled off the black cloth cover. Remaining motionless, he counted out twenty seconds and then jumped off the bicycle and quickly closed the negative shutter (at the bottom of the photo there is a slight shadow left by his feet as he dismounted).[3]

At about the same time, velocipedes began to appear in other Canadian towns. In Stratford, Ontario, it was reported that in March 1869 James Scarff, the mayor of neighbouring Woodstock, rode some twenty-seven miles over dirt roads to show off his velocipede in front of Stratford's City Hall. Among the impressed spectators was a local physician, Dr Robertson, who immediately purchased a boneshaker on which to make his house calls, and two enterprising Stratford mechanics, who set about making copies of the mayor's machine.[4]

These three anecdotes mark the beginning of a *carrier wave* that developed over the next thirty years. It culminated in something close to a mania for bicycles and bicycling in 1896 and 1897. Then, in the closing years of the century, it quite suddenly came to an end, as the trendsetters of Canadian society shifted their interest to new pastimes and the next wave of consumer luxuries and as the succeeding generation of young consumers identified with new manifestations of popular culture. In its turn, the bicycle carrier wave formed a small part of an even larger cultural movement known as *modernity*, which has its roots in the Enlightenment of the eighteenth century. Indeed, in western society the project of modernity formed a huge cultural movement that evolved over two centuries and, according to some commentators, today still has a powerful influence on western civilization.

Henry Poole, John Kerr, and Dr Robertson were therefore not simply early cyclists, but also players in a much larger cultural movement. Dr Robertson delivered modern medicine on a modern bicycle, thereby linking two of the most innovative thrusts of the movement at that time. The seventeen-year-old John Kerr brought together a pair of the most popular modern amusements of the late nineteenth century: photography and bicycling. Henry Poole's father, in shipping a bicycle from the imperial motherland to the booming resource town of Glace Bay, Nova Scotia, was diffusing a thoroughly modern consumer good and, in so doing, was anticipating a new and modern geography, which in the twentieth century has blossomed into the global economy. In different ways, all were caught up in the project to modernize daily life.

These early Canadian cyclists were therefore active during a particularly exciting period. The remarks of Keith Walden, writing about 'becoming modern in Toronto,' set the scene:

> The extent of change in the late Victorian Western world was staggering. None of it materialized abruptly out of thin air, nor was there any dramatic turning-point that marked the arrival of modernity, but by the closing years of the century even those most insulated from overt effects and most determined to resist intrusions could sense that Western society was shifting its axis. Cities grew inexorably; powerful new business organizations announced their presence with high-rise office towers and sprawling factories; bureaucratic centralization and regulation became more pronounced; migrations from distant parts of the globe snowballed; medical breakthroughs abounded; scientific discoveries spawned startling new theories; and bicycles, automobiles, dry-process photographs, halftones, movies, tractors, electric lights, typewriters, telephones, and other marvels of technological sophistication appeared with dizzying regularity ... The world seemed to have speeded up, to have become more complex. Change itself was in the saddle.[5]

How does the bicycle fit into this larger picture of rapid economic, social, and technological flux?[6] To examine this question, the bicycle is here set in the context of modernity at a specific time – the late Victorian period – and in a specific country: Canada. At first glance, this might seem an unlikely combination, since modernity has been the most powerful cultural movement in western civilization during the past 200 years, whereas the bicycle usually receives only passing mention in histories of this period. There are historians, however, who contend that, during the late Victorian age, bicycles contributed to social and economic change in several significant ways. Indeed, in his book on the modernization of rural

France from 1870 to 1914, Eugen Weber goes so far as to describe the bicycle as 'that incomparable harbinger of progress and emancipation.'[7]

The bicycle age corresponds with a phase when industry was central to the project of modernity. Industrial exhibitions had become very popular following the Great Exhibition, which was staged at the Crystal Palace in London in 1851. In the decades that followed, the great majority of major western cities sought to mount such expositions partly to confirm their importance as urban centres and partly because manufacturers were clamouring for opportunities to exhibit their newest machines and artefacts. For example, Keith Walden shows how the Toronto Industrial Exhibition, founded in 1879 and run annually until 1903 (when it was renamed the Canadian National Exhibition), had a powerful influence on a Canadian public eager to embrace a new urban culture composed of modern consumers. Among the many manufactured products displayed at these exhibitions were sewing machines, firearms, engines, industrial machinery, agricultural machinery, cameras, optical instruments, and bicycles. It is interesting that it was at the Philadelphia Centennial Exposition of 1876, where a number of British bicycle manufacturers were displaying their latest models, that a young American entrepreneur named Albert Pope was struck with the idea of manufacturing an American version. A year later he registered the trade name 'Columbia Bicycles,' and within twenty years the Pope Manufacturing Company was the world's largest bicycle manufacturer.[8]

The importance of context to modernity is illustrated by the setting in which the world's first true bicycles were developed and commercially produced, namely, Paris in the 1860s. If any city can lay claim to be the wellspring of modernity at that time, it is surely Paris. By the mid-nineteenth century, the celebrated French essayist Charles Baudelaire was urging his generation to observe the way modernity was transforming life on the streets of Paris. This invitation was subsequently taken up by many Parisians, most famously (among artists) by the impressionists, whose work T.J. Clark encapsulates as *The Painting of Modern Life*.[9] In literary circles, Emile Zola explored new aspects of Parisian life in his novels, notably in *Au Bonheur des Dames*, which is set in the world's first modern department store.[10] In effect, Paris was reconstructed by Louis Napoleon and the Baron Haussmann during the Second Empire (1852–70), as they tore down its medieval quarters and built the 'city of light.' It was on the newly macadamized boulevards of Paris, built at the behest of these urban developers, that Baudelaire saw modernity so vividly on display. It was also on these very boulevards and in nearby gymnasia and parks that the prototypes of the world's first commercially produced bicycles were tested.[11] During this period there was a palpable *jouissance* among Parisians for things modern.

Thus, it was no accident that the first wave of the bicycle boom coincided with Haussmann's obsessive efforts to build a modern city. The invention might have been made in London, New York, Boston, or even Vienna, but Parisians were arguably most in the thrall of modernity at that time. This geographical context is surely at least as important as the continuing debate over who was the actual inventor of the velocipede and when the invention took place. Although it was long thought to be the creation of Pierre and Ernest Michaux, David Herlihy has recently unearthed evidence that he interprets as pointing to Pierre Lallement as the true inventor.[12] Lallement claimed to have ridden a prototype velocipede he invented along the Boulevard Saint-Martin in Paris in the summer of 1863. This interpretation would seem to accord with the statement of Lacy Hillier, Britain's most knowledgeable cycle journalist of that period, who wrote:[13] 'The direct pedal action of a crank on the front wheel was first applied to a tricycle by Messrs Mehew of Chelsea, and the machine was shown in the Exhibition of '62. It seems probable it was seen there by an ingenious Frenchman named Pierre Lallement.' Michaux began commercial production of the velocipede in 1866, after which a velocipede craze erupted.[14] In 1868 Michaux opened a new factory, which was reported to employ 300 people and produce five velocipedes a day.[15] Several decades earlier, Karl von Drais, Denis Johnson, Louis Gompertz, Willard Sawyer, and others had invented proto-bicycles and other manumotive machines. Indeed, Hans Lessing and Roger Street report that the draisine had briefly stirred up quite a flurry of interest and for a short while was commercially produced in Germany and England.[16] But the draisine did not trigger a major innovation wave: within less than a year, interest in this early machine had waned. The velocipede craze of the late 1860s, in contrast, spread rapidly to other cities of the industrialized world, with the first reported sightings in Canada occurring in 1869. The initial boom in Paris was cut short following the onset of the Franco-Prussian War in 1870, but inventors, first in Britain and then in the United States, picked up on bicycle development where France had left off and the machine underwent rapid improvement for the next thirty years. In brief, the bicycle craze was launched in the city that had most consciously set about reinventing itself as a modern city, at a time when those efforts were in full spate.

THE PROJECT OF MODERNITY

The *project of modernity* was launched as an intellectual movement early in the eighteenth century by philosophers of the Enlightenment, including Berkeley, Locke, Hume, Kant, Montesquieu, Voltaire, and Descartes. Their purpose was to overthrow comprehensively the world view that had held sway throughout the pre-modern era, when the dogmas and practices of church and monarchy, the military, guilds, and landowners were imposed on the clergy, troops, peasants, serfs, apprentices, and other working people. In the decades that followed, challenges were made to the concentration of power in the hands of authorities that controlled various kinds of religious and secular hierarchies, and new, more democratic ways of governing society were contemplated. Challenges were also made to upholding belief and superstition and to the insistence on unquestioning acceptance of unchanging precepts. Such resistance to authorities was worked out locally, so that modernity found expression in many varied forms. Of course, pre-modern society and economy had not remained completely fossilized, but it was a society in which obedience was highly valued, dissent condemned, continuity stressed over change, and familiar ways were preferred to novelty.

The philosophical awakening of the early eighteenth century heralded a revolution in western intellectual thought that paved the way for modernity in its many diverse forms. According to certain interpretations, modernity provides a universally applicable and comprehensive understanding of subsequent history. There is, however, a competing conception of modernity that, by being situated and contextualized in particular settings of time and place, amounts not to one totalizing explanation, but to a cacophony of variations on the theme of modernity. As Miles Ogborn remarks: 'Modernity is most often a matter for grand theory and portentous pronouncements ... much of the theorizing about modernity ... presents it in ways that threaten to ride roughshod over the histories and geographies that could be written of modernity in different places at different times, ignoring the tricky issues of context, specificity, difference, and contingency.'[17] Philip Cooke, too, presents a geographically nuanced understanding of modernity, insisting that our experiences of modernity are transitory, fugitive, and contingent, not least because he sees 'the possibility of life being perceived as a project over which the individual has considerable influence, though not total control.'[18] In this study of the bicycle age in Canada, modernity is interpreted as a movement that is moulded and shaped in different ways in different settings, with difference emerging not only at the level of the individual, but also between groups, classes, city and countryside, regions, and enterprises.

This contextualized approach to modernity connects to two of the main versions of modernity identified by Ogborn. One version, characterized by a higher degree of universality, stresses the instrumental rationality that suffused the project. The emphasis on 'reason, rationality, and progress towards truth, beauty and the just life,'[19] was much discussed at that time in the salons of Europe's great cities. Out went the *ancien régime's* habit of looking backwards, and out went the mysteries of alchemy and superstition; instead, society began to look forward quite eagerly to the new and better things that rational thinking and scientific work would surely uncover in the future. Scientific and philosophical societies flourished in most British cities, and the cognoscenti took a growing interest in learning about 'the new.'[20] Despite the various groups who resisted the arguments of Enlightenment philosophers, including monarchists, traditionalists, and many more who had reason to fear change, a powerful group composed mainly of urban entrepreneurs took up the cause of building a more rational and progressive civilization.

The other primary vision of modernity emphasizes flux and change.[21]

Marshall Berman, in particular, dwells upon its mercurial qualities. Taken individually, many elements of this process of change are rational improvements, but taken collectively, they amount to an irrational and obsessive pursuit of change for its own sake. Although such a process is intrinsically creative, it is also necessarily destructive in that it requires a continual discarding of artefacts, practices, and institutions that are perfectly functional. Since changes within capitalist societies tend to be engineered by those in positions of power, they are not necessarily just, nor are they always reasonable. Indeed, given the nature of externalities, it seems difficult to imagine significant changes being made that do not have a negative impact on at least some sectors of society. Like the sword of Damocles that was suspended by a single hair, modernity is always poised to cut a swath across the established landscape.

Modernity was a complex and many-sided movement that at various times took on different aspects. Its origins were only partly philosophical. Paul Glennie and Nigel Thrift explore the early modern period, where they find evidence of modern consumption practices that pre-date industrialized mass production by about a century.[22] They suggest that in England, by around 1680–1700, fashion goods were beginning to appear. Manufactured by artisans, these goods were advertised, and regular changes of style were consciously introduced. They were widely traded and distributed via a network of smaller market towns, thus achieving considerable market penetration. The target was the emerging middle class, composed of comparatively wealthy farmers, traders, and artisans, who began to emulate the consumption norms of their more successful peers. By this time the population of British towns was growing rapidly – more so than on the Continent (except the Netherlands, where the population of cities was also booming). The identity of the emerging middle class became linked to changing patterns of consumption that included tobacco, tea, sugar, new fabrics, porcelain, clothes, furniture, and other new consumer goods.

By the latter part of the eighteenth century modernity had also developed a strong political dimension.[23] Indeed Gregory Baum characterizes modernity as 'the civilization initiated in the late eighteenth century by two major societal events, the Industrial Revolution and the democratic revolution.'[24] Baum argues that the experiments made during the American and the French revolutions to develop more rational and just forms of governance redefined our conception of democracy. Meanwhile, an era of profound industrial change was launched, although here the word *revolution* is misleading, since the pace of economic change picked up, if anything, during the nineteenth century. Industry played a key role in the advance of modernity, as an extraordinary succession of product and process innovations changed patterns of production and consumption beyond recognition.

By the mid-nineteenth century another expression of modernity had begun to occupy the minds of innovators: vigorous efforts were under way to design and build the modern city, especially in Paris. Soon thereafter, modernity took hold on the sensibilities of artists and writers, who began to paint and write about modern urban life. By the end of the nineteenth century efforts to create a modern city had turned to designing and building Utopian cities, among which Ebenezer Howard's garden cities are the best known. So the project continued into the twentieth century, with different facets of modernity receiving the most attention at different times.

The importance of the historical setting to the way modernity unfolded is illustrated by a vignette that relates to one of the earliest phases in the development of the bicycle. During the peace following the Napoleonic Wars the first proto-bicycles appeared as an amusement for (among others) decommissioned officers who were eager to impress young ladies. Baron Karl von Drais, a gentleman at the court of the Grand Duke of Baden, applied in 1817 for a patent on his draisine or hobby horse (the French used the term 'draisienne'). His invention appeared in that year on the streets of his home town, Karlsruhe, and on some of the surrounding country lanes. This draisine was propelled not by cranks or levers, but by the rider's legs, hence von Drais's name, 'laufmaschine' or running machine (see figure 1.2). In England it went by several names, including the pedestrian's accelerator, German hobby, or (most commonly) the velocipede, and was priced at from 8 to 10 guineas (approximately \$33 to \$42).[25] It is a measure of the spread of modern attitudes that the draisine

Figure 1.2. The draisine, invented in 1817 by Baron Karl von Drais de Sauerbronn (1785–1851). The inventor is shown in this 1819 painting by Joseph Paul Karg riding in Mannheim Castle Gardens. His forearms rest on a padded board, while he steers with a lever device attached to the front wheel. In the 1817–19 period, there was a brief craze for riding these machines in London, Paris, and several other western cities.

caught on very quickly, and similar machines were patented in the following year in London (by Johnson) and Paris (by Dineur, on behalf of von Drais).[26] The hobby horse was lighter than is commonly supposed and could do a fair turn of speed: Derek Roberts reports that on 12 June 1817 von Drais, himself, covered 14 kilometres along the Rhine Valley near Mannheim in under an hour,[27] while in 1818 a French rider made the longer ride of 28 kilometres from Beaune to Dijon, presumably on fairly rough roads, in under two and a half hours at an average speed of about eleven kilometres per hour.[28] The very fact that these speeds were recorded indicates how much these moderns were interested in going faster; the compulsion to erase old records and improve on old machines had taken a firm hold on their sensibilities.

This brief interest in the hobby horse also says something about values and attitudes as modernity gathered steam. At this stage, only the rich could afford to indulge in the sport. The innovation spread rapidly, however, making surprising geographical leaps, with reports, for instance, of the machines appearing in Calcutta, courtesy of the 'civilizing mission' of the British Raj, although I have not yet found any documents explicitly confirming that the hobby horse was seen in Canada.[29] By 1818, within months of their first appearance in Germany, hobby horses were being manufactured in London and Paris, and Americans followed suit soon after. Equally important, in less than a year, the new consumer had lost interest in the machine.[30] They went out of fashion almost as fast as they had come in. They were ephemeral. As Marshall Berman (1988) reminds us, many times, 'all that is solid melts into air.' Georgian consumers proceeded to embrace new sensations, including, by 1825, the first steam railways, leaving their hobby horses to gather dust in cellars and stables, with the occasional sortie as an amusing reminder of 'the old days.'

A reading of Canadian broadsheets available on microfilm suggests several key points about this early phase of modernity.[31] First, even at this time, there seems to have been a sequence of consumer luxury goods being marketed. The popular artefact of 1818 was the kaleidoscope; in 1819 the velocipede; while by 1820 microscopy was in fashion, with early microscopes focused on natural phenomena such as leaves of plants, wings of insects, and the capillaries of the human hand.[32] Second, as Bijker argues, it would seem that the mainstream of society has to be socially conditioned to accept an innovation, implying that an innovation can be ahead of its time.[33] In 1819, according to this argument, the bicycle had not yet entered the consciousness of a sufficiently large proportion of the population to trigger an innovation wave. Nevertheless, the hobby horse was frequently commented on in the press in the spring and summer of 1819. Although the concept of the bicycle aroused considerable interest in Canada and elsewhere, only a minority of citizens actually made a purchase (Roger Street reports that quite a number of them were wealthy aristocrats).[34] What is hard to assess is whether this reluctance was due to the high cost of a machine, to the large segments of society not yet socially conditioned to accept the innovation, or to the association of the machine by the summer of 1819 with 'dandies,' making it an object of fun and ridicule.[35] Third, the media (i.e., broadsheets) played a key role both in bringing the draisine to the public's attention and in helping it to fade away by ceasing to report on the subject. Long before Edward Herman and Noam Chomsky suggested it, the media were 'manufacturing consent' for things modern.[36]

In Baum's view, modernity also served as the whetstone of nineteenth-century politics, sharpening public opinion into three major movements, which, he suggests, eventually solidified into recognizable political movements.[37] Liberalism wholeheartedly endorsed the project of modernity and the various means, such as free trade, that gave it impetus. Conservatism, in contrast, lamented the arrival of modernity and disagreed with its democratic and technological assumptions. Radicalism, according to Baum, was not at odds with the idea of modernizing life, but took exception to the results; the benefits of modernity, in the radical view, were too unevenly distributed, and in the process a great deal of misery was created. This is, of course, a simplification of more complex political realities, but many of the proponents of the bicycle during the latter part of the nineteenth century were, in practice, loosely allied with the liberal faction, just as many contemporary advocates of the bicycle as an 'earth-friendly machine' are inclined to 'green' politics.

A key aspect of modernity was the degree to which it was manifested as a spectacle. Indeed, T.J. Clark sees this as the defining characteristic of modernity. He writes: 'the concepts of 'spectacle' and 'spectacular society' ... represent an effort to theorize the implications for capitalist society of the progressive shift within production towards the provision of consumer goods and services, and the accompanying 'colonization of everyday life' ... It points to a massive internal expansion of the capitalist market – the invasion and restructuring of whole areas of free time, private life, leisure, and personal expression ... It indicates a new phase of commodity production – the marketing, the making-into-commodities, of whole areas of social practice which had once been referred to casually as everyday life.'[38]

Figure 1.3. About thirty wheelmen parade down Queen Street in Saint John on the Elwell Bluenose Tour of 1886. Their captain, in front wearing a lighter coloured uniform, is riding the largest possible bicycle he could manage and can barely reach its pedals. A bugler would normally ride with such a parade to make sure everybody takes notice of this spectacle (see also figure 6.2).

From its inception, the bicycle had been a spectacle. One of the English nicknames for the draisine was a 'dandy-horse,' named for the regency dandies who made every effort to be as visible as possible. Little had changed when boneshakers were introduced half a century later; they appeared in the most visible places: the boulevards of Paris, the parks of London, and the avenues of New York. A decade later, members of high-bicycle clubs regularly put on a spectacle, riding in formation wearing military-style uniforms down the main streets of towns behind their captains, the club buglers noisily announcing their coming and signalling their manoeuvres. In figure 1.3 the progress of such a wheelmen's club down the main street of Saint John, New Brunswick, is shown. Some of the events mounted by wheelmen in the early phase of high-bicycle clubs echoed American Civil War practices, with riders arranged in cavalry formations, the calls of club buglers announcing manoeuvres, and club standard bearers leading the procession.[39] For the wealthier wheelmen these activities brought bicycling into everyday life; and for the masses, bicycle races became a popular entertainment (figure 1.4), with discreet gambling occurring on the sidelines. Six-day bicycle races were scheduled in huge tents, the crowd of spectators swelling during the midday break, and again in the evening. Bands piped up during peak viewing hours to entertain the crowd, and sprints were announced from time to time to add to the excitement. Even at small-town fall fairs, bicycle races were tagged on to the list of events to add some novelty to the spectacle offered to farmers, citizens, and their families. Throughout, bicycles were marketed in magazines and catalogues as commodities that defined the cutting edge of everyday life.

In the first years of this century Max Weber expressed his fear that modernity would lead to a society so rational, so scientific, and so technocratic that it would become like an 'iron cage' that snuffed out the creative side of life.[40] More recently, Marshall Berman has expressed similarly pessimistic sentiments about the turn of modernity in the twentieth century. He describes how, in New York City, the building of Robert Moses's Long Island Parkway in the 1920s, which liberated car-owning New Yorkers by providing them with easy access to the sands of Jones Beach, initiated a phase of destructive road building that culminated in the construction of the Cross-Bronx expressway, which in the late 1950s and early 1960s ripped the heart out of the old city. As New York was transformed, Berman sees Robert Moses mutating from a creative planner to a Procrustean tyrant.[41]

The destructive side of modernity is not new to the twentieth century. Haussmann may have been viewed as an enlightened planner when he was prefect of the department of Var, planning the shaded promenades of the southern French city of Draguignan. When summoned by Napoleon III to transform Paris and lead the conquest by the bourgeoisie of the contested spaces of the city centre, however, he gained the reputation of being

Figure 1.4. A bicycle race, part of a horse race meet at Peterborough's Morrow Park racetrack, circa 1885. High-bicycle races were a novel spectacle designed to add excitement to these events. Bets were often made on bicyclists as well as horses.

an authoritarian figure, much as Robert Moses did a century later. What distinguishes the recent phase of modernity is the level of technology that is available. A bicycle might spook a horse or injure a pedestrian, but it could not wreak havoc on the scale of late twentieth-century technology. Nineteenth-century technocrats were probably no less ambitious than their twentieth-century counterparts, but their actions were constrained by the tools available to them.

For geographers, modernity has taken on a particular meaning in relation to *place*. Just as technology has introduced instability into modern life – making it possible (and essential, within the terms of the project) to get rid of the old and replace it with the new, so places are now redefined and reinvented at regular intervals. Thus in the *modern* world, the character of a place is not tied to its natural surroundings, but is a creation of the various interests that 'produce' it or own it. Look, for example, at figure 1.5, a photograph taken in May 1899 in eastern Ontario as a cyclist disembarks from a train on the Carillon and Grenville railway. This region was settled by French and British farmers, who began to clear the land in the eighteenth century. With the arrival of the railway in the mid-nineteenth century, farming and forestry became more commercialized and rural folk found it easier to go to town, but the valley remained essentially rural. The bicycle subtly changed that state by making the countryside more accessible to the urbanite – here a bicyclist has taken a train into the country, from which point he will set out on a ride. The bicycle changed the meaning of the place called 'countryside,' producing a new space that included out-of-town hotels where cyclists could spend the night on a weekend ride, and tearooms for refreshment on shorter rides. In his book *The Betweenness of Place: Towards a Geography of Modernity* Nicholas Entrikin argues that place is 'contingent' in the modern world.[42] Places no longer have the stable meanings that they did over long periods during the pre-modern age. For instance, Vancouver's Stanley Park was a very different place for its bicycling clientele of the 1890s than it is today for its jogging and roller-blading aficionados. With the help of a continuous stream of new technologies linked to a succession of consumer fads, people define and re-define places. A place's identity is therefore contingent upon a series of planned and unplanned events that give meaning to it.

Figure 1.5. On 24 May 1899 a bicyclist disembarks from a train on the wide-gauge Carrillon and Grenville Railway in eastern Ontario to go for a ride in the countryside.

In some cases controversy surrounds the changing identity of a place such that it becomes a 'contested place.' It was always thus, but in the premodern era these meanings were much more stable. Modernity, of its very nature, seeks change and the status quo is rarely allowed to stand for long; change, indeed, may become the constant. The way bicycles and bicycling helped to give new meanings to certain sets of places and generated contests in others will be examined in the final chapter of this book.

It is also important to stress that modernity, in its various and complex forms, was not universally accepted. Throughout, there have been counter-currents and oppositional movements, in addition to groups, such as Old Order Amish and Mennonites, that have simply dissociated themselves from the project. In the formative stage of modernity, the judicial system of the Inquisition (which remained in place until the nineteenth century) threw up roadblocks to religious change wherever possible. The French monarchy did its utmost to stem the tide of political change. The Luddites did likewise for technological change, while the ecclesiastical establishment vehemently denounced Darwin's *Origin of Species*. Perhaps the most interesting of the many counter-movements was the romantic movement, spearheaded by a group of celebrated British poets who juxtaposed images of an oppressive urban-industrial society with descriptions of an idyllic rural life, innocent young countryfolk, and the virtues of a pre-modern existence. Romanticism diffused from its literary origins to influence nineteenth-century art and music – indeed, it may have been the most influential of all the nineteenth-century counter-movements.

CARRIER WAVES

For practitioners of modernity the paradigm of rational scientific progress was applicable to many fields of human endeavour, including engineering, science, transport, politics, medicine, communications, agriculture, and industry. These are fields where seminal inventions spawning a series of related innovations have ultimately had a much larger social and economic impact. Such a cluster of related innovations, if it achieves commercial success, is here labelled a *carrier wave*. For instance, the textbook example of the early steam engine was improved many times and linked with other innovations to be applied, subsequently, to mining, railways, steamships, factory production, and agriculture; in the process, the steam engine triggered a carrier wave that had a huge impact on industry and transport. In a similar manner, Barthelemy Thimonnier's primitive early sewing machine was improved many times, until it became a popular household appliance available with a range of accessories;[43] in addition, it changed the way the garment industry operated. Thomas Crapper's flushing toilet led to changes in house design, the construction of sewer networks beneath city streets and, eventually, to the building of urban sewage treatment facilities. The telephone has changed social relations, business relations, and, in turn, provided a network along which the contemporary knowledge economy is able to transmit vast quantities of information. The point is that the carrier wave concept, promulgated by Peter Hall and Paschal Preston to describe a specific late twentieth-century innovation wave based on information technology, has a far wider application. Moreover, it appears to be one of the more important mechanisms whereby modernity has affected contemporary civilization.

In introducing the carrier wave concept in the late 1980s, Hall and Preston did not say much about the mechanism itself, and they presented in their introduction only a brief dictionary definition of the term as it relates to electromagnetic waves. I believe it is a concept rich in potential and therefore offer an elaboration of the definition. Three criteria are suggested. First, the initial invention in a carrier wave does not remain an isolated breakthrough, but spawns a number of related innovations that improve upon or add to the original one, so that a connected series of innovations sustains the wave for some time. Second, the impact of the innovation does not remain confined to the original industry, but spreads through backward and forward linkages to related industries, often triggering innovations in these industries. Third, the main product of the carrier wave becomes attractive to consumers and therefore captures the public imagination – at least for a while – and in that sense has an impact on consumer culture.[44]

The carrier wave concept should not be read as a version of Whig history with technology rolling ineluctably onwards. There is nothing inevitable about the transformation of consumer innovations into major innovation waves; indeed, the great majority of inventions have no commercial impact whatsoever. Innovations that do succeed have had quite different impacts in different settings. One has only to look at geographical differences in the dominant use of the bicycle today: in much of Asia and Africa it is a utilitarian mode of transport; in France, Spain, and Italy cycle racing remains a major interest; in Britain and northern Europe cycle touring is more in fashion; while off-road mountain bikes have captured the imagination of many young North Americans. This, too, is a landscape painted with a broad brush. At the micro level subtle local geographies accompany large innovation waves, such as the development of bicycle courier services on the congested streets of Toronto, or commuting to work on the bicycle trail system of Helsinki.

While many innovations have evolved in unanticipated and serendipitous ways, others have lain fallow for years and then become important following the discovery of practical applications. A few terrifying innovations, such as nerve gases and biological warfare, in large measure have been suppressed. Finally, a small minority of seminal innovations have had major social and economic impacts that have transformed society.

All three criteria are amply satisfied by the case of the bicycle. It advanced technological modernity through a stream of innovations that subsequently proved crucial to the evolution of the automobile, the motorcycle, agricultural machinery, aviation, and many other forms of mechanical production, bringing in its train a host of related innovations. The methods used to make bicycles underwent dramatic changes as artisanal production was progressively replaced by mass production in vertically integrated factories. This transformation of production was an important precursor to twentieth-century mass production, particularly the version known as Fordism. It was matched on the consumption side by the bicycle's role as one of the industrialized world's first 'big-ticket, mass-produced, consumer luxuries' (utilitarian uses of the bicycle became dominant at a later date). Demand for bicycle products was stimulated not only by the succession of new bicycle models launched by manufacturers (each one designed to make the previous model obsolete), but also by the array of bicycle accessories advertised in magazines and brochures, which anticipated (a century later) consumer fads, such as the Barbie doll craze, that were similarly propelled by a wave of accessory sales.

Road improvement has a long history; the transition from bad roads to good roads began well before the twentieth century, when motorists launched their barrage of complaints about road quality. By the 1880s cyclists were grumbling about bad roads, and in the following decade they became active lobbyists in road improvement associations. More broadly, the bicycle had a significant impact on modernizing western society. High bicycles were ridden almost exclusively by men, with a macho element often in evidence (see figure 1.6, for instance). The safety bicycle, in contrast, was a gendered machine, with different configurations for men and women, while in the case of the tandem men and women could ride together. The bicycle thereby made at least a modest contribution to the emancipation of middle-class women and to the more informal fashions of clothing worn by both men and women. Most broadly, the bicycle

Figure 1.6. A race on the frozen Ottawa River between bicycles and skates. This engraving dates to 1880; the bicycles are early machines with straight handlebars.

helped to define new and modern places, such as the riding academy, the salesroom, the factory, the public park, and the racetrack, where bicycle-related activities took place. Of equal importance, the bicycle was geographically liberating; indeed, it produced new and often gendered spaces in the sense that Henri Lefebvre uses when he talks of the 'production of space.'[45] Bicyclists were able to go where they wished, at their own pace. This geographical liberation was taken much further in the era of the motor car, but modern understandings of the geography of personal space began to take shape during the bicycle era.

The proposed definition of a carrier wave adds some flesh to the concept, but the idea can be taken further, since it appears that waves typically evolve through a series of recognizable stages. The version presented here draws on several sets of ideas, including the product cycle model, innovation waves, and stage models of development. It is stressed, quite emphatically, that the concept is not intended to be a mechanistic model with universal applicability. What is claimed is that the concept captures a number of the major impacts of a successful innovation, including those on industrial production, on patterns of consumption, on corporate structure, on the provision of infrastructure, on social regulation, on social relations, and on geographies of individuals and of larger entities. This sequencing is quite deliberate, as will be explained. The following six components of a carrier wave are identified, each appearing roughly in the sequence they are listed.

The Innovation Phase

Important new products almost never appear in their final form. On the contrary, they usually go through a period of intensive experimentation and development. At times, especially at first, improvements may come so frequently that some versions of the product never go beyond being prototypes. This innovation phase is peppered with major technological insights that trigger sub-waves of related innovation. For instance, within the long series of bicycle-related inventions, John Boyd Dunlop's pneumatic tire, which he patented in 1888, subsequently triggered dozens of patents for valves, puncture-proof tires, and improved rims. Invention and innovation are likely to be focused on the main product at first, but at a later date may expand to embrace a series of related and ancillary products and accessories. One innovation may foster another; hence, inventions tend to be geographically and temporally clustered.

The Transformation of Production

Submissions for patents are often based on prototypes, built by their inventors. Many inventions – including some good ones – go no further or languish in workshops and attics for lack of capital to launch commercial production. Those that do enter into production usually follow the product cycle model beginning with small (often batch) production runs made in artisanal workshops with frequent changes of specifications.[46] This is followed by bigger production runs as the product's configuration stabilizes and demand grows. In due course, popular products achieve a standard form that permits mass production, thus lowering the unit cost, which encourages further purchases until the market is saturated. With a saturated market and declining demand, small-scale producers often are squeezed out, while the surviving larger firms consolidate to make the industry more oligopolistic.

The Transformation of Consumption

As the main product driving a carrier wave achieves public recognition, so entrepreneurs start to look for ways to embellish the product by developing, advertising, and selling related goods and services. The consumer is presented in due course with a wide range of accessories. The computer, for instance, now comes with a huge range of add-ons, as a visit to a large computer store soon reveals: colour printers and scanners, additional hard drives, extra RAM, restful screen savers, new software, and numerous other accessories. In some cases the combined cost of the add-ons may exceed that of the original purchase. The point to a carrier wave is that the artefact becomes a consumer good purchased not simply for its utility, but also as a social marker. In fact, many accessories and related purchases are designed to enhance the conspicuousness of the primary artefact and, therefore, of its owner. This is likely to happen not in the earliest, innovative phase, when the product may not yet be widely recognized, but later, when the artefact has become established as a signifier. Indeed, it may not be worthwhile to market an accessory until the product it serves has become sufficiently standardized that the accessory will work with a variety of models and generate a large enough market to warrant its development.

The Development of Related Social Overhead Capital

If the impression has been given that carrier waves represent the undiluted creative power of private sector initiatives, then that impression should be corrected. Industrial development is often assisted by government purchases, institutional support, training programs, the provision of public infrastructure, direct and indirect subsidies, and other public policies to assist the private sector. Of particular importance, as Gershuny and Miles note in their book, *The New Service Economy*, is social overhead capital that facilitates new forms of economic development.[47] Thus, the recent explosion of communications technology via electronic mail was made possible by the availability of sophisticated telephone and cable television networks. Elsewhere, the growth of electronic mail has created a strong lobby to improve telephone communications in regions where they are inadequate. Similarly, recent growth in the number and variety of television channels has been made possible by investment in public or private cable systems and in satellite technology. For bicyclists, the essential infrastructure was roads, especially because the poor state of many late nineteenth-century roads made bicycling a hazardous activity. As the bicycle carrier wave gathered momentum, so cyclists and cycle manufacturers responded to the dismal condition of roads by forming a vocal lobby to improve them and, in some instances, to construct bicycle paths. Social overhead capital usually becomes an issue only when a carrier wave has a sufficiently large following that an effective political lobby can be formed.

Social Regulation and Social Relations

In the innovative phase of a carrier wave, an artefact does not normally command enough attention to warrant much regulation. There may be some nuisance effects; cyclists, for instance, frequently had altercations with drivers of horse-drawn vehicles and were sometimes charged for riding on sidewalks and boardwalks reserved for pedestrians. Systematic regulation of an innovation usually is the result of there being broad social concern about its impact. Statutes and by-laws may then be promulgated in response to disputes and judgments in courts of law, while formal programs and policies may be the outcome of a protracted political process.

The social impacts of an innovation vary at different stages of a wave. At first, only the most adventurous may identify with an innovation; 'imitators' adopt it later as the artefact becomes more widely accepted. The major social impacts of an innovation take considerable time to negotiate, especially if it upsets the established order. However, the most important social relation of a carrier wave is its class relation. Successful innovations that generate carrier waves temporarily carry a social cachet and act as social markers. When an innovation becomes commonplace in the later stages of a carrier wave, it ceases to distinguish a privileged class of consumers and therefore rapidly loses its social significance.

New Geographies

The innovation generating a carrier wave can have immediate effects, but the larger geographical impacts may take longer to be negotiated. For instance, debate continues today over whether the long-term impact of electronic communications will be to disperse financial and corporate services to distant places, or whether the city core will retain its current hold on these contact-intensive jobs. With hindsight, the major geographical impacts of an innovation such as the railway may be quite apparent, but we are in a privileged position in knowing how history unfolded. At close range, it is often quite difficult to discern the long-term geographical impacts of an innovation. Moreover, since geographies tend to be culturally embedded, they change relatively slowly and are more likely to be accepted by a new generation than by older people set in their ways. In addition, since modernity is viewed as a contingent process, its geographical impact will depend upon the local setting, so that a carrier wave will produce several different new geographies.

Conceptualizing the bicycle era as a carrier wave helps to place it in a comparative setting. It draws attention to both similarities and differences in the way that various artefacts have affected society. It focuses interest, in places such as St Etienne and New England, on how the bicycle carrier wave connected to preceding phases of armaments and sewing-machine production. Equally interesting are its subsequent links to the automobile, aeroplane, and motorcycle industries in the British Midlands and the U.S. Midwest. Also, since geographical contingency is central to the impact of carrier waves, place-based studies draw attention to diversity and specificity of institutional, cultural, and physical forms. A brief overview of the bicycle age in Canada highlights a few of the distinguishing aspects of the Canadian version.

Figure 1.7. This engraving by Canada's foremost historical illustrator, C.W. Jefferys, captures the theme of this book better than any other illustration. The scene is set on Stephen Avenue (Eighth Avenue), Calgary, in 1902. Jefferys juxtaposes the traditional and the modern. Traditional frontier elements include the horses and the Blood brave in the foreground. The most visible sign of modern times is the bicycle, but careful inspection reveals several other elements, including electrification, street lighting, and new multi-storey buildings.

THE BICYCLE ERA IN CANADA

The choice of Canada as the place to explore the modernizing effects of the bicycle may seem perplexing, since, unlike Germany, France, Britain, and the United States, Canada was not one of the great hearths of bicycle innovation. This was not a country in which bicycling reached unprecedented popularity, nor was it a country that produced a string of celebrated bicycle racers. Yet despite not being at the heart of late-nineteenth-century bicycling, Canada, too, experienced a bicycle era. Quite simply, the contingent quality of modernity meant that the carrier wave

Figure 1.8. The velocipede rink and riding academy on Phoebe Street in Toronto in 1869 (the building on the left). The Protestant Orphans' Home is in the background. This pen and ink drawing is by William J. Thompson.

associated with cycling in Canada was experienced in a distinctive Canadian way (see figure 1.7), and within Canada a series of more local contingencies are apparent.

The Canadian bicycle era probably began in 1869, when the first velocipedes were seen; they enjoyed some success as riding academies made a brief appearance in Toronto, Montreal, and Halifax (figure 1.8). But this was a false dawn: boneshakers failed to attract a large following. Between 1870 and 1878 it was mainly British enthusiasts who carried on refining the bicycle, while Canadians temporarily lost interest. By 1878, however, ordinary bicycles (also known as highwheelers) were appearing more regularly on the streets of a few Canadian towns and the first club was formed in Montreal. These high bicycles, which were lighter, faster, and somewhat more comfortable than velocipedes, were exported from Britain to several eastern Canadian cities. Few novelties have attracted as much interest, and few were as visible. The front wheel of these bicycles was usually between 48 and 56 inches in diameter; the saddle was set above the big wheel on the back-bone, so that the rider was perched about 5 feet off the ground, at about the same height as a rider sitting on a horse. The term 'iron horse' has usually been applied to the railway engine, but it could be more appropriately used to describe the high bicycle. After all, like a horse, it was a form of personal conveyance, the rider was about the same height off the ground, and both were temperamental beasts quite capable of throwing their riders without warning.

How important were bicycles in Canada? Given the long, hard northern winters, when bicycles were usually put into storage, it might be expected that these machines would have had less impact than they had in countries with a milder climate. There were three counter-factors. First, Canada was a relatively affluent country. A basic bicycle cost $100 up to the mid-1890s, falling as low as $30 by 1900, when an excess supply of mass-produced bicycles was available. The average wage of a clerical or blue-collar worker was between $5 and $10 per week at this time, making bicycles too costly for most working-class Canadians, but they were not beyond the means of professionals, merchants, successful farmers, businessmen, and other citizens of the middle class. Second, the population included many young immigrants from the British Isles and European countries, people in the age group that took the greatest pleasure in bicycling. Third, Canadian urban centres were well developed by the 1890s. This is not to say that rural Canadians avoided the bicycle, but it was in towns that clubs were formed, races were staged, and cycle dealers were located. Although the weather closed down outdoor cycling for several months, bicycling clubs held a series of social activities during the winter, and journal reports suggest that bicyclists focused their social lives around bicycle club events for twelve months in the year. Besides, as we shall see, Canadians found a few ways to carry on bicycling, even through the winter months.

During their brief heyday in the late nineteenth century, bicycles arguably became the western world's number-one consumer luxury.

Demand was so great during the peak years that factories manufacturing goods as varied as agricultural machinery, clocks, carriages, and shovels, turned to making bicycles as a sideline. Owning a bicycle bestowed considerable kudos on its proprietor – indeed, high bicycles and tricycles, and some of the other two-wheeled machines that inventors dreamed up were among the most eye-catching artefacts yet produced. Bicycles and tricycles presented men and (later, when they became safer) women with a highly visible opportunity to demonstrate that they were in the mainstream of modern life.

In the late nineteenth century, owning a bicycle was a recognized indicator of modernity, just as having a web site is today. The great advantage for the cyclist was that the bicycle was used outside, under the public gaze, unlike many other consumer innovations that were used within the home, where they were less visible. This visibility gave bicyclists the advantage of immediate recognition (see figure 1.9); indeed, a case will be made in the penultimate chapter that Baudelaire's celebrated flâneur reappeared in the closing years of the century, but this time riding upon a bicycle.

At first, bicycles were expensive, and the boom in ownership was limited to people of high social status, such as those recorded in figure 1.10. Bicycle riding was therefore an activity confined to the affluent, many of whom formed exclusive clubs. Late in the bicycle era, when mass-production methods were adopted and supply exceeded demand, and when more second-hand machines were available, the cost of owning a bicycle fell dramatically. At the same time, wages earned by industrial and office workers were inching up, so that eventually a much larger number of consumers could afford to purchase one. The bicycle was a convenient and, by 1900, an inexpensive way of getting around, both for work and for pleasure, although Canadian winters made bicycling a seasonal affair for the great majority. The number of riders increased rapidly after 1890 as bicycles became much safer to ride and, with the addition of the pneumatic tire, more comfortable. They also became more affordable; for instance, the price of Columbia bicycles was halved between 1896 and 1900.[48]

The bicycle was, moreover, the object of numerous technological

Figure 1.9. A local photographer, Karl Lemp, recorded progress in his home town of Tavistock, Ontario, in about 1899. Here, a young employee of Maxwell's implement dealership displays his bicycle: this is an article that he wishes to be put on record.

innovations that gave impetus to sales, as carefully promoted new models were advertised to a receptive public. The form of the bicycle boom also reflected a key point stressed by Nicholas Oddy – that bicycles have been gendered machines almost from the start, with different configurations and models for men and women.[49] Faced with the prim and proper codes of mid-Victorian society, women were largely excluded from riding velocipedes and high bicycles, although a few rode tricycles. When, in 1881, Queen Victoria placed an order for James Starley's Salvo – thereafter Royal Salvo – tricycle (probably for her daughters to ride), the boundaries of acceptable female comportment were perceptibly pushed back. By the time of the safety bicycle, independent-minded young women began to see the bicycle as a means of asserting their independence, and redefining

Figure 1.10. Ottawa society gathers at the summer home of Mr E. Miall in Aylmer, Quebec, to be photographed by Mr E. Reardon of Ottawa. Nineteen of the latest pneumatic tyre safeties are lined up, probably at the start of the bicycle craze in the summer of 1895.

their identity with their own specific bicycles, costumes, and accessories. This trend was reinforced by Canadians such as Frances Willard, who advocated temperance and bicycling as the key to good health.[50] The newly opened market for women bicyclists heralded the peak of the bicycle boom, while at the same time playing a significant role in reshaping late Victorian society. Young couples also took to the bicycle, as shown in figure 1.11, and both in practice and in song ('A Bicycle Built for Two') attached a romantic meaning to the machine.

There was, in short, a crucial transition period between the railway age of the mid-nineteenth century and the age of the automobile that began early in the twentieth century. During this interval the bicycle was in vogue, indeed it gave rise to a boom in one of the first mass-produced leisure goods. This boom had an impact not only on economic affairs of production and consumption, but also on social life, culture, and even politics. It was a harbinger of many of the technical and social changes that manifested in the twentieth century. Yet few historians have attached much importance to the bicycle. For some it was an object of ridicule, especially the boneshaker and the ordinary or highwheel bicycle. Cartoonists poked fun at these machines (see, for example, figure 1.12), while riders of high bicycles frequently earned the wrath of horse drivers and pedestrians.

The bicycle also had to compete with streetcars (trams) for the attention of social historians of the late Victorian period. Both means of transportation became important during the same period, although their impacts were quite different. The streetcar has received considerable attention from urban historians, partly because it was capable of moving large numbers of people; bicyclists rode in ones and twos or in small groups. As shown in figure 1.13, both were an essential element of the Canadian street scene at the turn of the century, but they produced quite different spaces. Streetcars accelerated urban sprawl with the advent of streetcar suburbs, whereas the bicycle did not create bicycle suburbs.[51] At this time, a few Canadian cyclists were using machines for the journey to work or to shop, but they were used mainly for recreational purposes.

The main reason, however, that the bicycle has been overlooked by

Figure 1.11. A Columbia ladyfront tandem (the man seated at the rear can steer the machine). The young couple, W.K. Masterman of Montreal and his lady, pose in Notman's studio in July 1896. Of particular interest is the young woman's outfit. For this date, the skirt is unusually short, but she seems to be wearing some kind of leggings. Both riders have a bell. For courting couples, the tandem was a popular novelty.

Figure 1.12. This cartoon shows Sir Richard Cartwright, an advocate of reciprocity with the United States, riding a highwheeler with Sir John A. Macdonald perched on the small rear wheel (a fly on a wheel has exaggerated ideas of his own importance). This cartoon appeared in the humorous weekly, *Grip*, 19 May 1888.

historians is almost certainly what happened shortly afterward, when the North American and European love affair with the motor car began and continued with a passion that has since hardly abated.[52] In the intervening period the car has so completely mesmerized consumers, that other forms of ground transportation have been largely eclipsed. Only in the 1980s, amid the rumblings of the environmental movement, did attention in some quarters turn to Earth-friendly alternatives, such as walking,

Figure 1.13. Citizens on Yonge Street, Toronto, celebrating the fall of Pretoria in the South African War, 5 July 1900 – or possibly 30 May, since it was celebrated twice, the first time prematurely! According to an 1895 traffic census, 395 cyclists passed through the intersection of Yonge and King streets between 6:00 and 6:30 one summer evening.

light railways, and bicycles. At the beginning of the twentieth century, when the automobile was beginning to command the attention of Canadian consumers, the bicycle had ceased to be a defining social marker. Old bicycles were stored in the attic, hung up in the barn, or hauled to the town dump. They were no longer chic; they no longer carried the cachet of modernity. Nevertheless, during the boom years the bicycle had been instrumental in the march of modernity in several important ways that will be explored in subsequent chapters.

CHAPTER TWO

The Bicycle Carrier Wave

Tools, clothes, furniture, and bicycles can become old in two ways; they can wear out, and they can become obsolete, owing either to changes in style or to technological innovation. In pre-modern times, things became old mainly by use, whereas in modern times shifts in fashion and consumer tastes can make things old *before* they wear out.[1] Of its very nature, industrial modernity is connected with technological innovation. Such innovations are not necessarily 'good'; indeed, the heart of Berman's critique of modernity is that innovation has a Janus-like personality, with one face looking to make the world a healthier and safer place to live in, while the opposite face contemplates far more destructive outcomes.[2] Whether planned or unplanned, obsolescence has become the norm in modern society. Berman's mantra, 'all that is solid melts into air,' therefore, captures this sense of liquefaction of values and of styles that only a short time before had formed the very pillars of acceptability.

The bicycle captures several elements of industrial modernity. During the earlier part of the bicycle era, a number of the technological developments represented triumphs of ingenuity and imagination. The spider wheel, butted spokes, hollow rims, improved roller chains, and the axle differential that James Starley reinvented for his tricycles subsequently had important impacts on other forms of vehicular traffic. John Boyd Dunlop's reinvention of the pneumatic tire was another breakthrough of much greater general significance.[3] On the other hand, some of the inventions touted in brochures and letters of endorsement were as useful as snake oil.[4] There were impractical inventions that never made it off the drawing board, others that were manufactured but proved to be commercial failures, and a good number of innovations that sold well but were a success more because advertising had persuaded consumers that they were useful than because they actually represented significant progress. Since the subject here is bicycles in the mainstream, little attention is paid to the eyecatching but impractical inventions that are discussed more fully in other histories of bicycling.[5]

In this chapter the bicycle carrier wave is examined, especially as it evolved in Canada. Canada was not a primary innovation hearth for bicycles, but Canadians were nevertheless caught up, in their own distinctive way, in the development of bicycle technologies, in fabricating bicycles, and in cycling activities. Quite a number of British and American bicycle inventions were patented in Ottawa, presumably because their creators felt that their ideas had the potential to sell in Canada, and sometimes these inventions were adapted to Canadian needs. Conversely, a few Canadian innovations achieved recognition on the world stage. Thus, although Canada was a relatively minor player, there was a distinctive Canadian version of the bicycle game, as an examination of Canadian bicycle-related patents registered before 1900 demonstrates. As was the case elsewhere, the imagination of some Canadian bicycle patentees exceeded their grasp of practical mechanics, and, conversely, some talented Canadian inventors lacked either the commercial abilities or the capital resources needed to establish a business. Some of the claims made in advertisements almost certainly exaggerated the performance of bicycle innovations.[6] Moreover, towards the end of the era the majority of new patents offered only minor improvements to existing knowledge, but many still reflected the Canadian contexts in which the inventors operated.

LONG WAVES

Innovations to popular consumer goods such as the bicycle do not occur in isolation: they usually arrive in clusters. According to the neo-

Schumpeterian version of economic history, these clusters – such as those surrounding the internal combustion engine, radio communications, and the microchip – may create *long waves* of economic development lasting from fifty to sixty years.[7] Historians have attached slightly different dates to these waves. The schema followed by Dunford and Perrons and Hall and Preston identifies four long waves, each associated with a group of key propulsive industries, as follows:[8]

1787–1845	First Kondratieff	- cotton, iron, steam
1846–1895	Second Kondratieff	- railways, steel, machine tools, ships
1896–1947	Third Kondratieff	- autos, electrical goods, chemicals
1948–2000	Fourth Kondratieff	- electronics, computers, communications, aerospace

There is some difference of opinion among economic historians over the key innovations driving each wave, but none lists the bicycle as a key propulsive industry. In this chapter, the bicycle will be put back in the picture, since it is here viewed as a minor carrier wave, not important enough to define a long wave, but nevertheless providing a significant bridge between the second and third Kondratieff. Evidence from the bicycle era suggests that innovations that bridge two long waves are conceptually important, even though they are not stressed in the extant literature. Propulsive industries driving each long wave do not emerge from the ether, but develop in a technological sense out of preceding discoveries and innovations and in a social sense out of new perceptions of the practical applications of recent innovations.[9]

Long-wave theory presents a somewhat totalizing view of recent economic history, with successions of innovations driving industrial modernity ineluctably forwards. This comprehensiveness is both a strength and a weakness. Its strength lies in its presentation of a unified and seamless narrative of economic development since the industrial revolution. The very comprehensiveness of that narrative has three important shortcomings, however, each of which has relevance for the bicycle boom.

First, the long-wave model focuses on big innovations and pays less attention to the smaller innovations that may pave the way for major innovations or connect one major innovation to another. In practice, the bicycle falls between the second and the third Kondratieff, connecting innovations in machine tools, armaments, carriages, railway, and steel production of the earlier wave with the automobile and machinery industries of the third Kondratieff. Indeed, the automobile age might not have evolved without the foundation of a series of trailblazing bicycle innovations introduced during the previous Kondratieff. Early motorcycles and aircraft also benefited from bicycle-related innovations.

Second, long-wave theory assumes considerable uniformity in the way innovations spread, whereas in reality, biased communication and trade channels and local cultural idiosyncrasies created very particular geographies of innovation diffusion. For instance, Britain exported bicycles to its dominions scattered around the world, but it sought to exclude bicycles made in other countries from these markets. Canada, having established a bicycle industry, exported bicycles mainly within the empire. Early high bicycles sold in the United States were imported mainly from Britain, but once an American industry was established, British bicycles lost their U.S. market share. In Japan and Russia there was very little bicycle activity at this stage, but bicycle manufacturing began in Sweden, Germany, Czechoslovakia, and Italy. Within Canada, bicycles were of minor importance west of Winnipeg, where settlement on a large scale began only at the end of the bicycle era; the main innovation hearths were cities in eastern Canada.[10]

Third, the long-wave model pre-dates recent thinking about the local embeddeness of innovations. In practice, bicycle manufacturing developed in a relatively small number of towns and regions, several of which had a tradition of metalworking and engineering, including, in most cases, the manufacture of armaments and/or sewing machines. The leading bicycle manufacturing centres in Britain were Coventry, Wolverhampton, and Birmingham; in France, St Etienne and Paris; and in the United States, New England, where Hounshell specifically connects the bicycle industry with the New England (Yankee) armament-manufacturing tradition.[11] Moreover, bicycle innovations took on a different character in different places. In France, innovation appeared early, as Lallement, Michaux, and Cadot developed the velocipede.[12] During the Franco-Prussian War of 1870–71 the French bicycle industry switched to armament production, while the British bicycle industry flourished. In Britain, innovation was focused more upon improving the machine itself, whereas in the United States in the 1880s and 1890s greater attention was paid to innovation in production methods. In 1879 the American bicycle entrepreneur, Albert Pope, subcontracted the production of his first batch of bicycles.[13] In less than fifteen years Pope was the largest manufacturer of bicycles in the world, and in the 1890s he pioneered some of the mass-production methods that Henry Ford later took much further in his automobile factories.[14] The concept of embeddedness makes the Schumpeterian approach to innovation more geographically relevant while simultaneously challenging the universal assumptions implicit in long-wave theory. Thus, for example, we find French bicycles and advertising posters that reveal a flair for style, whereas functional aspects were more valued in Britain. The French and Italians took more interest in racing bicycles than North Americans did. By the turn of the century the Dutch were making reliable and practical roadsters, while later American design concentrated on comfortable, but heavy 'sidewalk' bicycles, ridden locally and not over long distances.

EMBEDDEDNESS

To suggest that innovation is an embedded process is to suggest that the context of inventive activity matters. It does not follow that genius and insight play no role in the process, but the probability of making a technological breakthrough is increased by access to good information channels and by the local availability of those with various related skills, such as persons who can help to build a prototype or a specialized test apparatus. In short, innovation is seen as a social as well as a scientific process.

The concept of embeddedness can be traced back to the work of the great anthropologist, Karl Polanyi, who used the term to capture the idea that in non-market economies, economic relations – buying, selling, producing, and consuming – are 'embedded and enmeshed' in social relations, and (implicitly) in the historical and geographical contexts in which these social relations have developed.[15] Thus, for example, large tributes may be paid to a temple or a ruler, which in market terms makes no sense. In 1985 the concept of embeddedness was given new life by Mark Granovetter, who recognized that, even in market economies, social conventions and institutional practices are important to economic life.[16] Granovetter connected the concept of embeddedness explicitly to the branch of institutional economics that stresses transaction costs. To oversimplify, it costs time and money to do a deal, but if deals are repeated (and become embedded) in dense and stable social networks where actors come to know and trust each other, then deals cost much less to negotiate. For this reason, most economic transactions become socially embedded. Granovetter's use of the term 'embedded' is evidently more specific than that of Polanyi.

In the decade since Granovetter's paper was published, the concept of embeddedness has been increasingly used by anthropologists, geographers, and sociologists. The meaning attached to the term has shifted yet again, however, to emphasize the way local and regional cultures influence the operation of economies. Culture, in this sense, is an umbrella term that includes all sorts of established local practices, the formal and informal rules that govern the functioning of institutions such as guilds and trade associations, the modus operandi of networks of artisans who specialize in various activities, the system of apprenticeship that reproduces particular skills in succeeding generations, and the trading practices used to distribute the final products to markets.

This notion of embeddedness helps us to understand the record of bicycle patents in Canada in several ways. First, the process was socially embedded, with patents registered in three waves. The third wave, which was by far the largest, corresponded with the bicycle craze of the mid-1890s when a kind of herd instinct operated. This period of intense social interest in bicycling was accompanied by a flood of patents on improvements to the machine and its many accessories. Note that by 1900 the interests of movers and shakers in society had moved on to other things, so that, although more bicycles than ever before were in use in Canada, far fewer bicycle patents were registered. When the interests of trend-setters in Canada moved on, so did the socially embedded process of technological innovation.

Bicycle-related patents were also geographically embedded. Certain regions and towns, to be identified presently, accounted for a disproportionate number of patents, whereas other regions were scarcely involved. A key factor in this variable geography was the local presence of industries that fostered skills easily adapted to bicycle manufacture. Rather more specific to Canada was an environment that influenced the innovation process both through the good supply of suitable wood available at low cost (wood being a raw material used in the 1890s in bicycle construction) and via the challenge of the long winters, when Canadians were looking for new outdoor activities to amuse themselves. Finally, certain groups of people – whom we might label 'avant-garde' – took an interest in improving the bicycle, while other groups took no interest at all.

It is proposed, therefore, that the large-scale innovation processes described by the carrier wave mechanism are, in practice, modified by the way they are locally embedded. Some local economies were rather hostile to bicycle innovation, while others were clearly quite supportive. Resource and farming communities, for instance, were less given to mechanical invention than the towns of central Canada engaged in making various types of machinery.

THE EVOLUTION OF THE BICYCLE, 1868–1900

The major elements in the technological progression of the bicycle are found in four key types of bicycle, each of which has an appropriate descriptive name: the boneshaker (or velocipede); the highwheeler (or ordinary bicycle); the hard-tire safety bicycle; and the pneumatic-tire safety bicycle. The focus on these 'mainstream' bicycles should be explained, since almost every bicycle book that has ever been written waxes eloquent about the hundreds of fascinating and often bizarre machines that were invented during this period. For the most part, the evolution of the bicycle was not defined by these failures, although some – like Henry Lawson's 'bicyclette,' which had a chain drive to the rear wheel – were crucial to later developments. Some good ideas emerged too late; for instance, the American Eagle of 1889 – a high bicycle with the small wheel at the front to prevent falling forward over the handlebars – was introduced after the chain-driven safety bicycle had begun to achieve widespread acceptance. Dozens of inventions were totally impractical and were doomed from the start. Some were good ideas, but their inventors lacked the resources to promote and sell the product. Industrial modernity was edging towards mass production, which could be achieved only if the herd instinct among consumers created sufficient demand to cover the heavy cost of investment in mass-production techniques. Note, also, that as far as can be said with certainty, only the last of these four types of bicycle – the pneumatic-tire safety – was manufactured in Canada.[17] For sure, Canadian blacksmiths and wheelwrights made copies of the boneshaker and the highwheeler, but as far as is known, they did not manufacture them in numbers.

In figure 2.1 a boneshaker manufactured by Wood Brothers of New York is shown. The manufacturer's serial number is 64, suggesting that it was made early in the velocipede craze, which was shorter lived in the United States than in Europe – 1868 is its likely date of manufacture. The spokes and felloes are made of hickory, ringed with an iron tire, but the spokes do not project from the hub in a straight line (as in a wagon wheel) but are set alternately on opposite sides of the hub to create a triangular cross-section for greater strength. The hub is also of hickory, with a 1-inch-square steel axle centred through it. The front forks and main strut, which are made of solid iron castings, are extremely heavy. The saddle is attached to a leaf spring that is 33 inches long and quite springy so that the rider yo-yos up and down over potholes. The saddle consists of a metal pan padded with horsehair and wrapped with two pieces of shaped leather sewn together along the edge of the pan. The saddle sits on a wooden chock that lifts it off the spring and is bolted in place. The bearings follow the model of railway bearings, being composed of two split brass fittings that form a collar around the axle. The pedals are brass cylinders that rotate on a steel axle; riders place the arch of their feet on the pedals. Finally, steering is done by straight handlebars, which also activate the brake by rotating the handlebars to tighten the twine that pulls a metal

Figure 2.1. The boneshaker. These velocipedes were usually painted brightly, often with bold stripes. The front wheel of this machine measures 35 inches in diameter, the rear wheel 29 inches. Riding this bicycle on a modern paved road is a bone-shaking experience, but on dirt roads, such as were the norm in the late 1860s, the ride is actually quite smooth.

brake pad onto the rear iron tire. The evolution of the velocipede out of wagon and railway technology is quite evident.

The highwheeler, or high bicycle, or ordinary bicycle, or penny farthing, evolved from the boneshaker.[18] Recent research suggests that this happened very quickly (between April and October 1870), when it became apparent that ordinary bicycles beat velocipedes in bicycle races at Aston Cross Grounds in Birmingham and Molineux Grounds in Wolverhampton.[19] The inventor of the 'spider' suspension wheel fitted to the ordinary was Eugene Meyer of Paris, although most of the subsequent development of the ordinary bicycle occurred in Britain.[20] First, a digression on names, since bicycle historians have spilled a great deal of ink on the name of this bicycle. As Derek Roberts stresses, at first it was known simply as the bicycle, because there was no other bicycle, except the slow and clumsy boneshaker, which had gone out of fashion.[21] The need to distinguish this bicycle from other bicycles arose only when other types of bicycle attracted a large following, as happened after 1887. The telling comparison is with the safety bicycle, because high bicycles were unsafe. Then, as now, riders of high bicycles were prone to doing a 'header,' that is, pitching forwards after hitting a pothole, breaking a spoke, having a tire come off, or touching the pedals at high speed. Since the rider's legs were trapped behind the handlebars, he (very rarely, she) planted his face in the dirt, and if he tried to break the fall with his arms, two broken wrists were a common result.

In figures 2.2 and 2.3 two stages in the evolution of the high bicycle are shown. The first was probably manufactured in England in the early 1880s, about the mid-point in the evolution of the highwheeler. The different sizes of the front and rear wheels, already evident with the velocipede, is now taken to extremes. The pedals directly drive the big front wheel, which vary in diameter from 48 up to 60 inches. The length of a rider's legs was the main limitation on how big a wheel could be ridden. The larger the wheel, the faster a rider could go, but safety considerations might result in the choice of a somewhat smaller-diameter wheel. For the young men of the 1870s and 1880s, as much as for those of today, speed was of the essence; hence, a competitive rider would purchase a bicycle with the largest-diameter wheel his legs would permit. The rear wheel is as small as possible (typically with a diameter of 16 to 18 inches), since it was dead weight for a rider in a hurry. The bicycle illustrated in figure 2.2 has direct (straight) spoking, whereas the bicycle in figure 2.3 (circa 1887) has tangential spokes that cross over each other several times. Tangential spokes make a wheel stronger by spreading the force of jolts and, according to some riders, make for a smoother ride.[22] In figure 2.3 the rim of the wheel is no longer wood, but hollow metal (to reduce weight). The tire is made of hard rubber and is glued or cemented on (today, they are usually held on by wire). Such tires are not very shock absorbent. The backbone is a hollow hard steel tube, cold-swaged to a taper, so that it has a bigger diameter at the top than at the bottom. Sometimes the backbone was oval,

Figure 2.2. An ordinary bicycle made in the early 1880s. With a front wheel probably measuring 54 inches diameter, it is capable of a top speed exceeding twenty miles per hour. In the absence of reliable bicycle chains and gearing, this was the design that afforded maximum speed. The rear wheel simply balanced the machine; the smaller it was, the less it weighed. This photo, taken in Notman's Montreal studio, is of Mr A. Harris, wearing the uniform of the Montreal Bicycle Club. This machine has two interesting accessories: a Hill and Tolman bell and a hub lamp suspended in the front hub.

Figure 2.3. An ordinary bicycle made circa 1887. There are a number of differences from the machine shown in figure 2.2: the spokes are tangential, the saddle is a sprung hammock (and quite comfortable), the handlebars are moustache shaped to allow the rider's legs to rise higher at the top of the stroke, and the grips are spade shaped, which relieves tension on the wrists. The rider is Mr E.L. Laliberté, one of the first French Canadians to gain prominence in a sport that had previously been dominated by the Anglo-Canadian elite. Several racing medals are pinned on his uniform. Notice the leather toolbag buckled in the coils of the saddle spring.

for extra strength, but backbones more commonly were round. The forks are also hollow for lightness and strength, although the very earliest high bicycles had solid forks. The saddle shown in figure 2.2 is a metal pan, not dissimilar to that of the boneshaker, but soon after, more comfortable hammock saddles, such as that shown in figure 2.3, appeared, especially on American high bicycles. A small step is placed on the backbone, about 18 to 21 inches off the ground, to help the rider to mount the saddle. The brake is normally a spoon apparatus activated by a lever on the handlebars, although on early high bicycles the brake was sometimes on the rear wheel. It is not a very effective brake; if it were, it might lock on the tire at speed and cause a header.

Bicycle manufacturers experimented with layouts aimed at improving the safety of the high bicycle through the late 1870s and 1880s. Broadly speaking, there were two main strategies. First, there were bicycles that continued with the high-bicycle format but sought to make them safer. These included: (1) lever-driven high bicycles, such as the Xtraordinary and the Facile, which moved the saddle and pedals (and the rider's centre of gravity) backwards with respect to the front axle; (2) bicycles on which the small wheel was switched to the front, such as the Star and the Eagle; (3) bicycles that used gearing (with a chain on the Kangaroo, or hub gearing on a Crypto) to reduce the size of the large wheel; and (4) bicycles on which the centre of gravity was moved as far backward as practical, such as the Rational. The second strategy, which turned out to be the successful one in the long run, was the safety bicycle with two wheels of approximately the same size driven by a chain to the *rear* wheel. The first to be manufactured were solid-tire safety bicycles. Compared with modern bicycles, these were heavy machines and were laborious to ride; indeed, at first they were slower than high bicycles, which explains why there was a period in the late 1880s and early 1890s when both types of bicycle often appear in photos (figure 2.4). In 1890 and 1891, for instance, the Toronto Bicycle Club had two first lieutenants, one for ordinaries and one for safeties. Thin-walled metal tubing was not yet used for frames, and chains were quite heavy, typically being made of 1-inch blocks, which are less flexible than the modern half-inch roller chain.

Figure 2.4. The transition from high bicycles to solid-tire safeties. Uniformed members of the Hamilton Bicycle Club pose on the steps of Hamilton Public Library in 1888. The bicycle in front of the library's doors is an early cross-frame hard-tire safety. The remaining twenty-four bicycles are ordinaries. The navy club uniforms are highly decorative. Note the club bugler (front left), the bracing of pairs of bicycles, and the captain (mounted) centre. Three years later, another photo of the same club shows all but a handful of riders on safety bicycles.

Figure 2.5 is a studio portrait of D.M. Laughlin and his solid-tire safety bicycle. This an interesting early safety, photographed in 1889. The main support of the cross frame (which was later superseded by the diamond frame) runs from a hinge attached to the steering head back and down to the rear hub. At this time frames were made of either solid steel or fairly heavy tubing, which, with heavy cranks and spokes and solid rubber tires, made these machines exceedingly ponderous. The seat tube is hidden by the rider's leg, but it is probably S-shaped, with two wire stays connecting the seat tube with the steering head to give the bicycle stability. On the front fork, note the springs that absorb vibrations and the footrests for coasting down hills (this bicycle has a fixed gear, so its rider gets a rest only when his feet are on the coasting pegs).

John Boyd Dunlop's pneumatic tire, which he patented in the United Kingdom in 1888, was adopted fairly speedily, even though at first it was unreliable and easily punctured. It is reported that at the Stanley Bicycle Club Show held in the Crystal Palace, London, in November-December 1891, 40 per cent of the bicycles exhibited were fitted with pneumatic tires, 32.5 per cent with cushion tires, and 16.5 per cent with solid tires; the remaining 11 per cent had tires that were 'inflated [*sic*] or nondescript.'[23] By 1893 the pneumatic tire was fitted to the great majority of bicycles, and from then on the hard tire and cushion tire rapidly fell out of favour. Meanwhile, the safety bicycle was steadily improved. Thin-walled tubes, silver soldered or brazed together, made light, rigid frames. Rims also became lighter; indeed, wooden rims had become very popular in Canada and the United States by the early 1890s, whereas most European bicycles were fitted with metal rims. Meanwhile, women were becoming enthusiastic bicyclists, creating a growing demand for drop-frame bicycles without cross-bars. Women riders soon began to wear skirts a couple of inches above their ankles to avoid getting their feet caught up in the hem as they rotated the pedals. They also quickly adopted chain guards and skirt guards on the back wheel to prevent their clothes from getting caught up in the moving parts. Brakes began to improve, and chains became lighter, stronger, and more flexible, thus reducing the amount of resistance they created.

Figure 2.5. A solid-tire cross-frame safety bicycle: Mr D.M. Laughlin of Montreal, mounted on his hard-tire safety in June 1889. Note his toolbag, revolving bell, long socks and elaborately tailored jacket. The coasting pegs on the front fork give his legs a rest in downhill sections. The front forks are sprung just below the coasting pegs. The cross-frame was heavy, but not very strong.

CITIUS, ALTIUS, FORTIUS

The Olympic motto, 'Citius, altius, fortius,' urging athletes to run faster, jump higher, and perform feats of strength, has both a classical and a modernist ring to it. This motto was proposed by Father Didon, principal of Arcueil College in France in 1895, who taught in a highly competitive education system, which had evolved from the French Revolution. The motto could be applied equally to the technical development of bicycles during the same period. Mechanics and inventors certainly competed to make ordinary and safety bicycles go faster, and many were as obsessed with distance and speed records as car racers were to be in later years. They proclaimed the strength of their machines in advertisements with large numbers of bulbous men precariously loaded onto a single-seater bicycle or standing on its frame. Higher? To stretch a point, high bicycles used for racing were built with as big a wheel as possible, given its rider's leg length; indeed, the saddle was usually lowered and clamped directly onto the backbone to permit a larger wheel. The Giraffe safety bicycle had a very high frame, thereby giving the rider a high centre of mass, which, according to Sharp, increased its lateral stability and steadiness in steering as well as making the rider more visible![24]

The technical development of the bicycle between 1868 and 1900 can be summarized under five major elements: the frame, the suspension, the wheel, the transmission and steering, and the brakes.

The Frame

Forging and casting were the methods used to make the solid and heavy metal frame of the boneshaker yet despite this weight, it was quite whippy and the rider had to adapt to the flexing of the entire machine . By the 1870s improvements in manufacturing technology and design made frames lighter, stronger, and more rigid. The switch to steel tubing for the backbones and forks of highwheelers in the 1870s therefore represented a major breakthrough, because it reduced the weight of the frame very substantially with no loss of strength and some gain in rigidity. Better swaging technology made the tapered backbones of ordinaries lighter and stronger; tubing at the top of the backbone typically had a diameter of about

1.5 inches, tapering to around 1 inch where it joined the rear forks. Being made of hard steel and swaged, the backbone was, in effect, a large spring.

Early safety bicycles made in the 1880s had fairly thick-walled tubes. Following development of the process of butting seamless lightweight tubes in Britain in 1887, Reynolds tubing came into widespread use on quality bicycles in the 1890s. Meanwhile, Mannesmann in Germany developed a thin-walled tubing. The Pope Manufacturing Company, manufacturer of Columbia bicycles at Hartford, Connecticut, purchased the rights to use Mannesmann's patent in the United States, and in 1892 the firm began to mass produce thin-walled seamless steel tubes in Hartford for Columbia bicycles and for Gormully and Jefferies of Chicago. Most bicycle manufacturers had followed suit by the mid-1890s. Both the cross frame (see figure 2.5) and the diamond frame were used with the earliest safety bicycles. Sharp shows how the cross-frame design concentrates forces and a bending effect on the point where the two main tubes cross.[25] The forces are spread more evenly on the diamond frame, which became the norm by the early 1890s for men's bicycles. The combination of thin-walled tubing and the diamond frame completed the main elements of frame evolution, creating essentially the same frame configuration in use today for road bicycles. Distinctive women's bicycles appeared with the advent of the safety bicycle; their frames with no top tube allowed a woman wearing long skirts to mount easily. The frames were notably weaker, however, particularly those with a single down tube, which put considerable stress on the join to the bottom bracket.

The Suspension

A person riding a boneshaker on a paved road today experiences a lot of vibration, which is passed from the road surface through the frame. The softer dirt roads of the 1860s were much more forgiving, however, and the experience was far less uncomfortable than present-day riders of these machines imagine. Attached to the backbone of most models of boneshakers is a long leaf-spring that flexes up and down over bumps, lifting the rider out of and down into the metal pan saddle.

The suspension of the ordinary bicycle brought several improvements. The hard rubber tire of the ordinary absorbed some of the high-frequency rattle, while the larger wheel also smoothed out minor undulations.[26] The switch from direct to tangent spoking in the 1880s may also have improved matters slightly. The early pan saddles of ordinaries were no better than those of boneshakers, but when the leather was tensed across the saddle or replaced by a hammock, the ride was significantly improved. Also of importance, the backbone was a long springy tube, which damped out some of the vibration coming from the small rear wheel.

Interest in suspension reached a peak during the hard-tire safety era. Some hard-tire safeties were equipped with spring suspension on the front fork, as shown in figure 2.5. The rear end was not normally sprung, although sprung-saddle posts absorbed some of the shocks. Once the shock-absorbing pneumatic tire was in widespread use, the problem of vibration was much reduced, although, with increasing numbers of women riding, there was renewed interest in the mid-1890s in creating a smoother ride. Springs were still sometimes attached to the front forks and the handlebars were occasionally sprung, but more effort now went into springing the back part of the frame. The 'Hygienic' rear spring, which was quite widely used on Canadian safety bicycles after 1900, was placed below the saddle and above the rear forks. Sprung saddles also appeared in the pneumatic safety era, adding yet more cushioning against vibrations.

The Wheels

Bicycle inventors also played a key role in developing modern wheel technology. The hubs of boneshakers were made of wood (hickory, ash, or oak), usually with a 1-inch-square steel axle centred through it. The protruding parts of the axle were machined round and set in a split brass collar clamped together by a cotter pin and lubricated by grease or oil to form a bearing. Early ordinary bicycles also had collar bearings with oilers fitted to them, and their riders carried oilers with them at all times to relubricate the bearings when they ran dry. Conical bearings also were used on early high bicycles, having the advantage that they could be adjusted inwards to take up slack as they wore.[27] These early highwheelers were equipped with a brass or a steel hub. Roller bearings or ball-bearings did not make an appearance until around 1879. Dodge describes how Daniel

Rudge, who had just patented his adjustable ball-bearing axle, showed it to the celebrated French bicycle racer, Charles Terront, in 1879.[28] Having failed to persuade the makers of his bicycles to fit the new bearings, Terront did it himself and immediately set speed records on the Paris Longchamp racetrack. It was the Bown 'Aeolus' bearing, however, that was most widely used on highwheelers and tricycles up to the mid-1880s. After the Bown patent had expired, a range of alternative ball-bearings began to appear.[29] The evolution of bearings that were light, strong, and offered very little resistance to a rotary motion was crucial not only to bicycles, but to all forms of modern land transport.[30]

Hubs are connected to felloes and rims by spokes. At first, spokes were made of wood (hickory was preferred), which made them quite light and strong, but prone to warping. A key innovation, patented in 1869 by W.F. Reynolds and J.H. Mays, was the suspension wire spoke used on their 'Phantom' bicycle, an important transitional machine between the boneshaker and the highwheeler. Its steel spokes were quite heavy, but they did not warp ; the main problem was tensing them. The first solution, developed and patented by James Starley and William Hillman and used on the Ariel of 1871, was to rotate the hub using a tension lever until the wheel spokes were taut. By the mid 1870s spokes were threaded, so that they could be screwed into matching holes in the hub until they became tense and the wheel 'true' in three dimensions. Early metal spokes usually formed a radial pattern (the spider wheel) so that as the wheel rotated the weight of the rider was transferred to the spoke directly above the hub. This system, known as direct spoking, was largely superseded around 1885 by tangential spokes, where the spokes criss-crossed each other, so that the weight of the rider was spread by several spokes to an arc at the top of the rim.

In the last two decades of the century, better steel-making technology made it possible to manufacture lighter and stronger spokes. Butting, where the spoke is thicker at each end and thinner in the middle to reduce weight and wind resistance, became common. In the late 1880s the adoption of threaded spokes attached to the rim with nipples and a mushroom or ball at the hub end of the spoke made wheel trueing much simpler; this system also allowed the size and weight of hubs to be reduced

Figure 2.6. Fixing a puncture on the Tour of Tours ride from Hamilton to Niagara Falls in May 1896.

substantially. Rims followed a similar pattern of evolution, beginning with the wooden felloes of the boneshaker and followed by the metal rim of the high bicycles. The profiles changed from rather heavy crescent-shaped rims to rolled hollow rims as steel-rolling technology improved.

Over a period of twenty-five years, the tire arguably changed more dramatically than any other part of a bicycle. The metal tire of the boneshaker was carried over to many blacksmith-made high bicycles, although it is of interest that a few boneshakers were fitted with solid-rubber tires. Better-quality rubber made the switch to solid-rubber tires an obvious choice for ordinary bicycles in the 1870s. As noted earlier, Dunlop rediscovered the inflatable tire in 1888, but the next tire innovation, widely used in 1890 and 1891, was the cushion tire, which was lighter, fatter, and softer and gave a smoother ride than the solid-rubber tire. The problem with the cushion tire was that it deformed to the shape of the road, which created considerable resistance: a smoother ride was achieved at considerable extra effort! Inflatable tires that held their air and did not puncture every few minutes achieved general acceptance from 1892 onward, although, as figure 2.6 shows, fixing punctures was a ritual that all serious bicyclists had to endure.

The Transmission and Steering

A bicycle's transmission connects the feet of the rider with the rotation of the wheels. Arguably, the most important commercial innovation occurred quite early when cranks were attached to a wheel.[31] Prior to this, the presumed but, as Dodds concludes, 'not proven' bicycles of Kirkpatrick Macmillan and Gavin Dalzell apparently were driven by treadles and levers similar to those used on industrial machines, spinning wheels, and railway engines of that period.[32] Attaching a crank and a pedal to an axle was a fundamental breakthrough and remains the essence of bicycle propulsion to this day. The resulting rotational movement is the most efficient form of propulsion yet discovered; it has never been matched by levers, ratchets, pulleys, or any other mode of propulsion. Albert Pope's genius was to spot the importance of Lallement's crank patent and, by moving swiftly, persuade its various owners to sell it to him for an incredibly small amount.[33] Pope then reduced the crank patent fee from $25 to $10 and made a small fortune. Since he wanted to popularize the bicycle, he did not wish to discourage sales by imposing a prohibitive patent fee.

The development of pedals was more modest. Boneshaker pedals usually were of metal, on a greased bearing, with the rider placing the arch of the foot against the pedal. The pedals of ordinaries were not greatly different from those widely used today: flat, with two rubber bars and a ball-bearing and attached to a spindle. Moreover, early in the 1870s, racers of high bicycles developed pedal clips to keep their feet from slipping off.

Little changed in the fundamentals of transmission between the boneshaker and the ordinary bicycle. Both used direct rotation of the front wheel, the gearing being governed by the diameter of the wheel. The safety bicycle brought a major change, however, since the pedals were connected to the rear driving wheel by a chain. In wrapping around at least two sprockets, a chain creates the possibility of gearing up or down. A large chain wheel connected to a smaller rear sprocket raises the gear ratio and allows a smaller wheel of perhaps 28-inch diameter to achieve the gearing equivalent to a 60- or 70-inch wheel, or an even larger one if desired. For example, the Racycle Pacemaker (manufactured early in the late nineteenth and early twentieth centuries in Miami, Ohio, and in Berlin, Ontario) had a huge chain wheel with 40 teeth for a 1-inch pitch roller chain, and the rear sprocket had either 9, 10, or 11 teeth attached to a 28-inch wheel; hence, for the 9-tooth sprocket the gearing was 28 x 40/9 = 124.4 inches!

The ability to change gears was the next breakthrough. Gears were inserted in rear hubs, so that by 1900 bicycles with two or more gears were available, allowing a rider to choose lower or higher gearing depending on the terrain or whether she or he felt energetic. At first, gear changing had to be done while the bicycle was stationary, but by the late 1890s gears could be changed while the bicycle was in motion.

An interesting exception to chain technology was the chainless, or shaft-drive bicycle that had been available in Europe since the early 1890s. Pope's Columbia version, driven by a shaft connecting two bevelled gears,[34] was launched with great fanfare in October 1897. The shaft-drive bicycle sold well from 1898 to 1900, when it was a widely promoted novelty. Cleveland, for instance, manufactured a chainless bicycle at its Toronto factory. The shaft-drive mechanism prevented skirts from becoming caught in chains, but it created considerable friction; a skirt guard and a chain guard on a chain-driven bicycle achieved the same result more effectively. The chainless bicycle illustrates the promotion of a novelty that was not, in practice, an improvement over available technology.

Another area in which technology developed dramatically during the bicycle era was the steering and power drive of tricycles. For the most part steering was effected by attaching handlebars directly to the wheel to be steered, although indirect methods, such as bridle-rod steering, also were found on some machines. Of particular interest was the rack-and-pinion steering used on hayfork and loop-frame tricycles and the Coventry Rotary; it was subsequently adapted to various automobiles. Of even greater importance to transmission was another breakthrough developed for the tricycle by James Starley; then called the 'balance gear,' it was a differential on the driving axle that allowed power to be transmitted to both wheels (previously, only one wheel had been a driving wheel). Now a standard feature of all vehicles, the differential allowed the parallel wheels of a tricycle to turn at different speeds as it cornered.

The Brakes

Bicycle brakes became increasingly sophisticated during the boom. The most important brake on most early bicycles was provided by the pedals. Since these bicycles had fixed wheels, pushing against the rotation of the pedals slowed the machine and gave the rider some control when descending a hill or stopping. Boneshakers were also fitted with metal brake clamps adapted from those fitted to wagons. Brakes on the high-wheeler were never a priority; indeed, then as now, quite a number were ridden without the brake attachment, since a powerful brake on a high bicycle increases the danger of a header. With the introduction of safety bicycles, however, brake technology developed. By 1900 a variety of brakes was available, including plunger brakes that pushed down on the front wheel; lever brakes that pulled rubber brake pads on to the rim; coaster brakes activated by pedalling backwards; and disk brakes.[35]

Bicycle technology developed rapidly during the period 1868–1900. The 1868 boneshaker and the 1900 safety bicycle are as different as chalk and cheese. The boneshaker was an extension of centuries-old wagon and carriage technology, whereas manufacture of the 1900 safety bicycle required sophisticated metallurgy, milling, stamping and pressing, electric welding, precision gearing, high-quality rubber, case-hardened steel bearings, nickel plating, and exact and interchangeable parts. In the intervening years, the bicycle had been the focus of a great deal of inventive activity, which had given rise to a large number of practical inventions, many of which were registered as patents. An examination of bicycle-related patents registered in Canada provides evidence of this activity.

CANADIAN BICYCLE PATENTS

France, Britain, and the United States were the three most important bicycle innovation hearths. By the late nineteenth century, however, consumer fashions were spreading quite rapidly among the industrialized nations, anticipating the globalization of consumption patterns that occurred in the late twentieth century. The case for viewing the bicycle boom as evidence of incipient globalization is strengthened by the fact that even in a somewhat peripheral country such as Canada, consumers picked up on trends very soon after they were developed in the main innovation hearths of Europe and the United States, albeit in a contingent way marked by particular Canadian twists and turns. The Canadian case is of interest, partly because many of the patents registered in Canada were of foreign origin (reflecting this tendency to globalize), but mainly because of the way Canadian bicycle patents, and bicycle activities as a whole, were embedded in various ways in Canadian economy and society.

It was two years from Confederation in 1867 before the dominion's legislators managed to pass the Patent Act of 1869; prior to that date, patents of the new dominion were still registered and numbered under the old patent series dating back to June 1824. The new series, beginning with patent #1 of August 1869, at first applied to the four provinces of Nova Scotia, New Brunswick, Quebec, and Ontario. Other provinces came under the aegis of this act as they joined the dominion.

Bicycle-related patents registered in Canada between 1868 and 1900 are recorded by year in figure 2.7. These data encapsulate the bicycle era, with the frequency of patents falling into three phases. The boneshaker caused the smallest flurry of interest between 1868 and 1870. After that small ripple, virtually nothing happened until 1880, when a second small wave began; a series of patents were registered through the high-bicycle era, tailing off at the end of the 1880s. On the appearance of the safety bicycle, a third and vastly bigger surge of innovations occurred, peaking at 379 patents registered in 1897. Thereafter, interest declined rapidly, so that by 1900 only one-quarter of the 1897 tally was registered. Thus, there were three clear phases of interest, each bigger than the preceding one, broadly corresponding to the boneshaker, the highwheeler, and the safety bicycle.

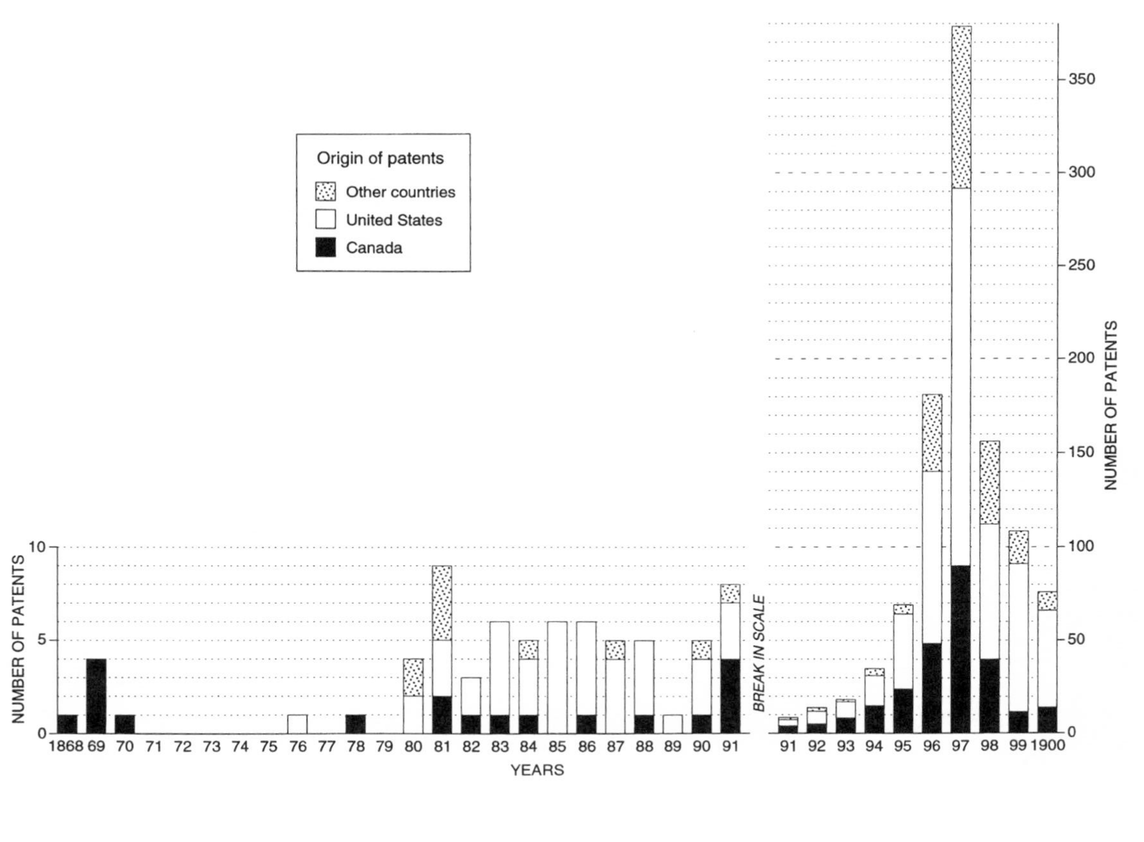

Figure 2.7. Bicycle-related patents registered at the Canadian Patent Office, 1868–1900.

The country of origin of these patents is also revealing. At first, they were exclusively Canadian, since inventors were required to live in Canada for twelve months before applying. This residency requirement was removed in 1872, and thereafter patents could be registered by citizens of any country. During Canada's high-bicycle period from 1880 to 1890 a total of fifty-five patents were registered, of which only eight (15 per cent) originated in Canada. It seems that since high bicycles were not manufactured in Canada, there was little interest among manufacturers in advancing technology. Seventy per cent of the patents in the high-bicycle phase originated in the United States. Until the mid-1890s most of the registered inventors were individuals rather than corporations, although a number of these inventors worked either with or for major bicycle manufacturers, including Lucius D. Copeland, who patented his steam tricycle in Ottawa in 1888, and William S. Kelley of Smithville, New Jersey, who in 1886, patented the ratchet system used on the Star bicycle.[36]

During the third phase, from 1890 to 1900, a number of Canadian firms began to manufacture safety bicycles, and a substantial stream of Canadian bicycle inventions were registered: in most years, between one-quarter and one-third of the patents were of Canadian origin. About half of the remaining patents originated in the United States, and the records of all other countries pale by comparison. A handful of British patents were registered in Ottawa, including several by John Boyd Dunlop and Edward C.F. Otto, but most European inventors did not bother to register their ideas, since Canada was seen as a minor and distant market. It is of interest, however, that quite a number of Australian and New Zealand patents were registered in Canada during the 1890s; bicycling had become popular in these dominions, as it had in Canada, and several manufacturers had been established.[37] Massey-Harris of Toronto and the Goold (Red Bird) Bicycle Company of Brantford developed a significant market in Australia during this period.[38] The year 1897 marked the peak of inventive activity: in this one year alone were registered almost one-quarter of all the Canadian bicycle-related patents recorded between 1868 and 1900. After 1897 there was a sharp decline as the interest of the leisure class in cycling waned rapidly. This historical clustering of patents reflects the social embeddedness of invention and innovation.

There is no denying that the pioneers of cycling, including Baron von Drais, Pierre Lallement, James Starley, and their ilk, were independent-minded technical leaders. Such pioneers needed to be somewhat iconoclastic loners in order to pursue their idiosyncratic visions in the face of considerable scepticism and some ridicule from other segments of society. In later years the great majority of inventors belonged to a very different crowd, who were clearly surfing a popular wave. Thus, the peaks of innovation waves reflect the flow of popular taste and fashion.

Being the year when patent activity peaked, 1897 is also a good year to use as an illustration of the global geography of Canadian-registered patents. Ninety (or 24 per cent) of the 1897 patents were of Canadian origin, compared with 202 (or 53 per cent) that were American. The remaining 87 patents originated in eleven other countries, including 36 in England; 3 each in Scotland and Ireland; 11 in Germany; an astonishing 19 and 10 in Australia and New Zealand, respectively; and 1 each in Austria, Belgium, South Africa, the Netherlands, and Switzerland. Clearly, the bicycle carrier wave swept the industrialized world, but it took a different form in each country.[39] In Canada, three component waves can be identified.

The First Wave: 1868–1879

The initial wave of Canadian patents consisted of improvements to the velocipede and the tricycle plus (as early as 1869) an ice bicycle. Compared with the third wave, in this period there was a very low level of innovative activity, mainly because the velocipede was not made in Canadian factories, although a few were made by artisans. Two examples capture the main innovative thrust of the period.

The very first bicycle patent submitted in Canada – Stimson's velocipede – was dated 24 October 1868 (#2910 First Series) (figure 2.8a). It was the creation of a medical doctor, James Stimson, who resided in St George, near Brantford, Ontario. His velocipede had no pedals: the rider rocked backwards and forwards in the seat, and this movement activated a spring-and-lever mechanism overhead, which in turn drove the sprocket wheel attached to the rear wheel with an action rather like riding a camel. Despite being Canada's first bicycle patent, this submission did

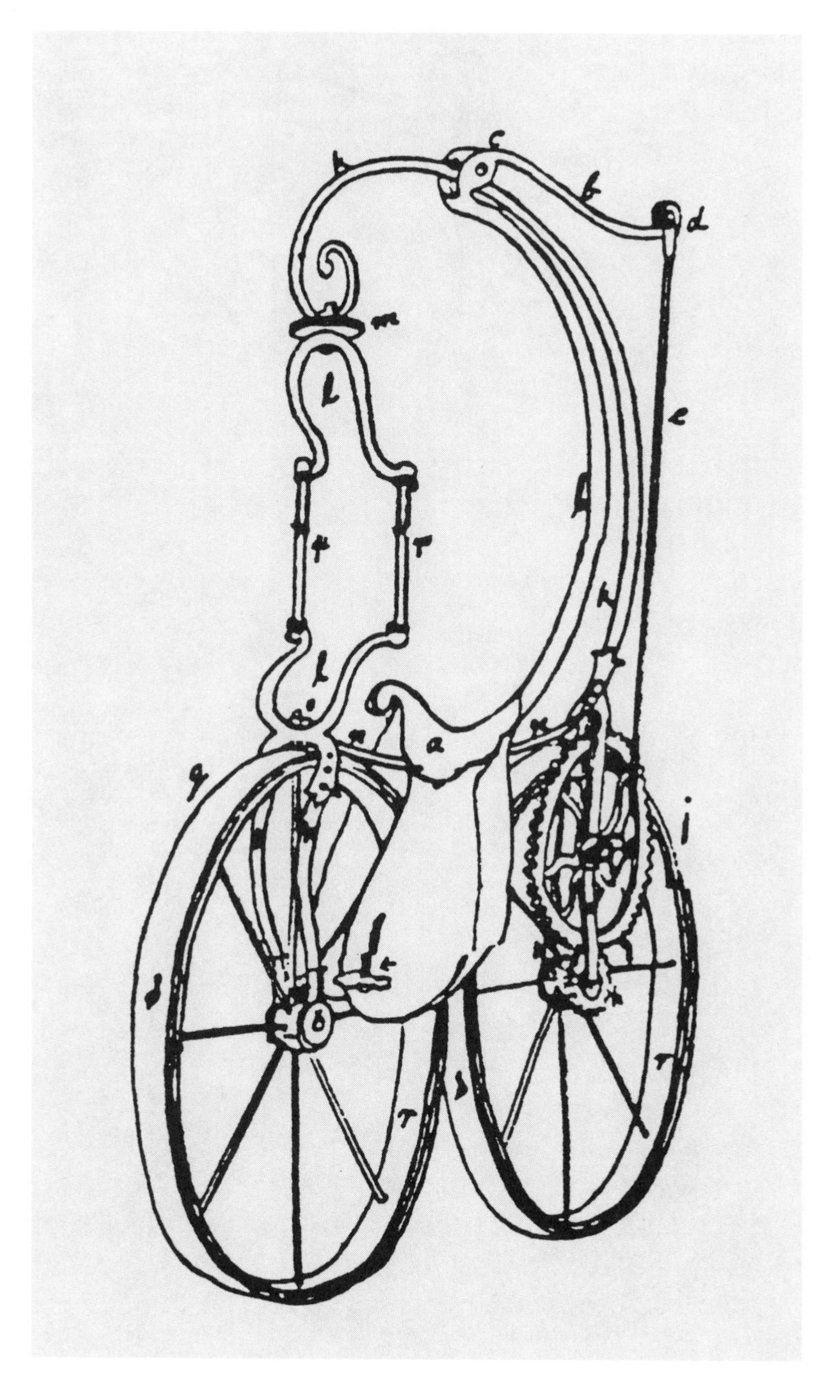

Figure 2.8. Selected Canadian bicycle patents: (a) Stimson's velocipede, 1868.

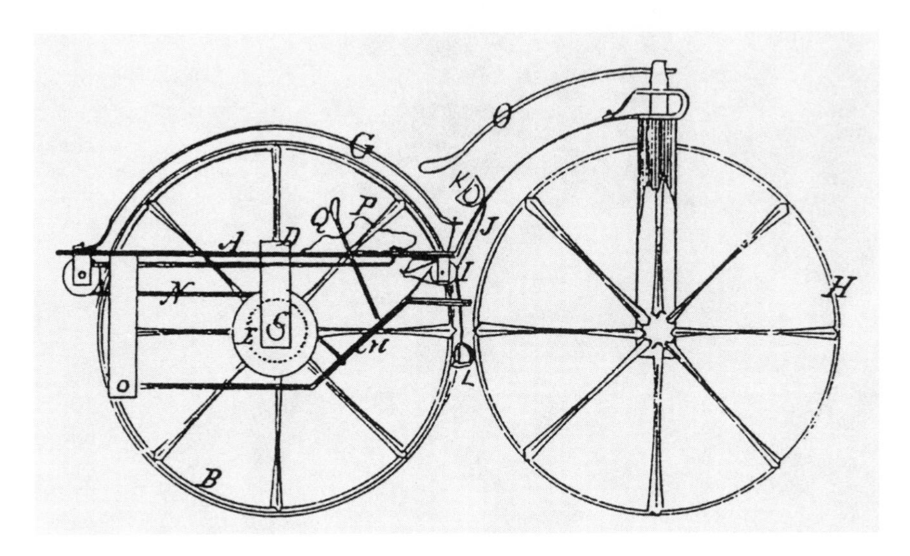

Lateral section.

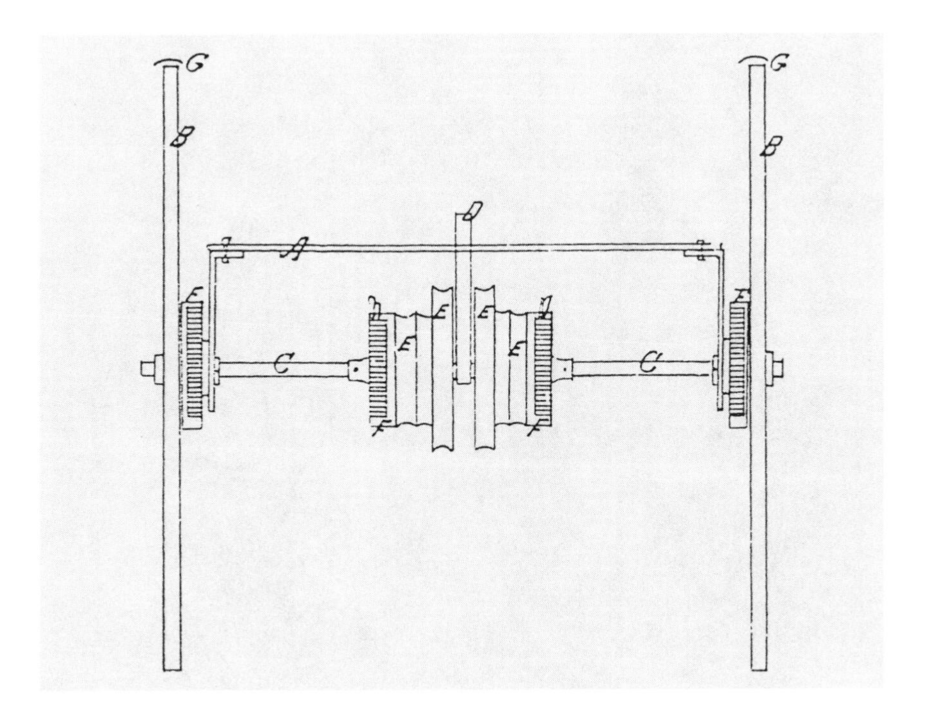

Cross section through driving wheels.

Figure 2.8. (b) Mason's improved tricycle, 1878.

not come from nowhere. Newspaper reports about the velocipede in the summer of 1868 aroused considerable interest, as did the first appearance of the machine on Canadian streets. Stimson must have thought about the design of the velocipede and presumably felt he could do better. As a doctor, he could afford to have a working model made, and, living near Brantford, he had access to the skills and technology used by the local agricultural machinery industry.[40]

The roots of Mason's improved tricycle of 1878 (#8191) lie in nineteenth-century industrial machinery (figure 2.8b). Allen Mason was a 'mechanician' in the small saw-milling and furniture-making town of Paisley in Bruce County, Ontario, working on a daily basis with the drive shafts, pulleys, levers, and ratchets on which his improvement was based. The diagram Mason submitted is not very informative, but it seems that the rider stood between the two rear wheels of a tricycle with his feet attached to driving stirrups, while his hands pulled on two levers that drove the tricycle, using a ratchet mechanism, so that 'every pound of human weight, in combination with human muscular force ... was used.' The machine had a split rear axle, each side controlled by a separate ratchet, and two gears, a low gear for hill climbing, and a higher gear for speeding.

The Second Wave: 1880–1890

Like the first wave, this period was marked by a low level of domestic innovation, whereas non-Canadians, including Edward Otto, Lucius Copeland, and John B. Dunlop, were active in registering patents in Ottawa. Most of the patents registered during this phase by Canadians relate to velocipedes and tricycles, where solid construction took precedence over speed. The first Canadian submission that used technology related to the ordinary bicycle was Robinson's tandem ordinary of 1884 (figure 2.8c).[41] Two machines are selected from this second wave for illustrative purposes.

Bailey and Thorne's Improved Tricycle of 1883 (#17,851) originated in Patterson, now one of the ghost towns of Ontario (figure 2.8d).In 1883 Patterson was a thriving community, where a large agricultural implements manufacturer, Peter Patterson & Bros, had made the decision to

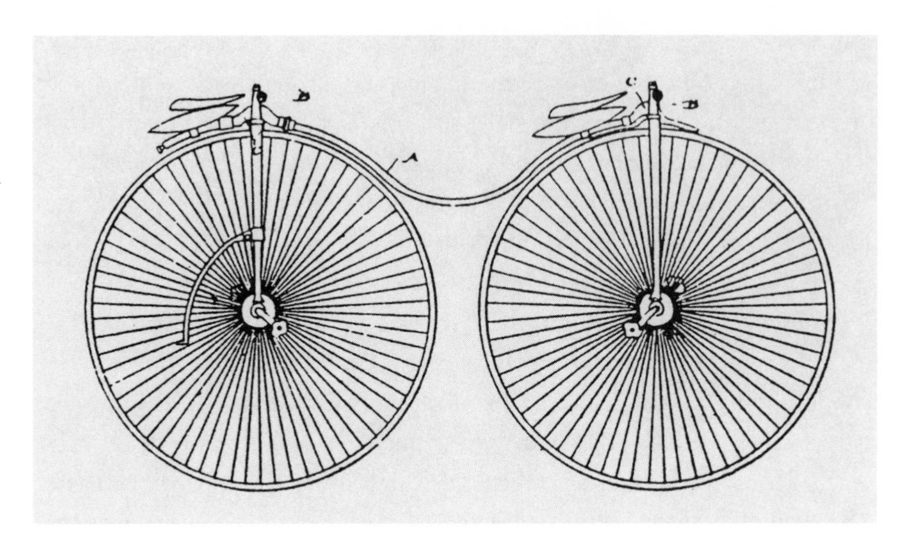

Figure 2.8. (c) Robinson's tandem ordinary, 1884.

Figure 2.8. (d) Bailey and Thorne's improved tricycle, 1883.

Figure 2.8. (e) Rourk's geared highwheeler, 1888.

diversify its product range into tricycles. The essence of Bailey's patent was a bent rear axle driven by stirrups attached to a jointed frame. This innovation illustrates how technology moved between related industries – in this case between agricultural machinery and bicycles.[42] It seems likely that this tricycle entered into limited commercial production.

A geared highwheeler patented in 1888 (#28,835) was the brainchild of Francis Rourk of London, Ontario (figure 2.8e). To activate this highwheeler the rider depressed a lever that turned the large sprocket wheel about one-quarter of a revolution, driving the bicycle forwards. Pawls fitted to the axle sprocket allowed the spring-loaded lever to recover for another stroke. Placing the lever behind the front axle reduced the likelihood of a header, so this bicycle was a safety highwheeler, its closest relative being the Springfield Roadster. London, Ontario, was a city with a diverse manufacturing base and one of the most active cycling communities in Ontario during the 1880s. In short, both the interest and the know-how were locally available to Francis Rourk, although his machine was not commercially produced.

The Third Wave: 1891–1900

During this wave a considerable number of Canadian firms began to manufacture bicycles. A few firms were established specifically with this objective, but the majority were engaged in related activities, such as manufacturing agricultural machinery, clocks, prams, and carriages; they began to manufacture bicycles in the hope of adding a profitable new line to their business. Previously noted streams of innovation, such as ice velocipedes and tricycles, continued to appear in the patent books during the third wave, but many of the patents concern two important new thrusts, the first being improvements to the safety bicycle and the second – particularly after 1895 – being bicycle accessories.

Improvements to the Safety Bicycle

In 1890 the safety bicycle was still a ponderous machine with (in Canada) mainly solid rubber tires, thick-walled tubing, and heavy spokes and rims. In the years that followed, numerous improvements were made to the safety bicycle, so that, by the end of the decade, a much lighter and more flexible machine was available with thin-walled tubing, pneumatic tires, and finer spokes, often set in light wooden rims. Hub and coaster brakes were replacing plunger brakes, and lighter roller chains were superseding heavier block chains. The bicycle became a light, fast, and robust machine.

Also in this period, newly established Canadian bicycle manufacturers began to register patents. In 1891 the Goold Bicycle Company of Brantford registered a patent on a safety bicycle, and thereafter Thomas Fane and Charles Lavender (of the Comet Cycle Company in Toronto), Peter Gendron (of the Gendron Bicycle Company in Toronto), Albert P. Jones (of the Toronto Cycle Company), Henry Cutler (of the Wanderer Cycle Company in Toronto), the Massey-Harris Bicycle Company (also in Toronto), the Clarksburg Wood Rim Company, and the Welland Vale

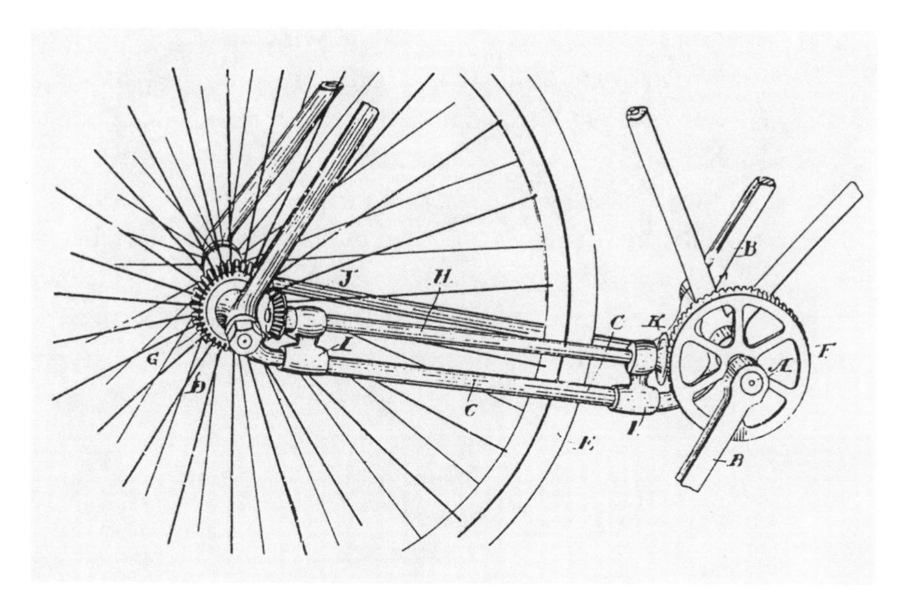

Figure 2.8. (f) Cutler's shaft-drive bicycle, 1893

Manufacturing Company (in St Catharines) submitted one or more patents based on their improvements to the safety bicycle.

During the first half of the decade, many of the innovations were mechanical in character, including various drive mechanisms and gears, flywheels, cranks, pedals, a shaft drive, plus sociable and tandem bicycles. By mid-decade, attachments to the bicycle – saddles, handlebars, mudguards, carriers, pumps, brakes, and chains – became more prominent. There was also something of an obsession with the search for a puncture-proof pneumatic tire. Synopses of three interesting innovations will give some idea of the range of contributions made by Canadians during this phase.

Henry Cutler's shaft-drive bicycle of 1893 (#42,414), which was based on a bevelled-gear shaft drive, was patented shortly after the Metropole shaft-drive bicycle appeared in France and five years before the Pope Manufacturing Company launched its successful range of chainless bicycles (figure 2.8f). As a manufacturer in Toronto, Cutler had access to a wide range of technical expertise. Cutler's patent is important in that he introduced gearing to this bicycle by fitting two driving wheels of different size. This was an invention ahead of its time, but its inventor evidently lacked the resources to produce and market it. The most successful chainless bicycle, the Columbia, was introduced quite late, but it was superbly promoted by the Pope Manufacturing Company.

Banes and Cleland, two Toronto-based mechanics, patented a pneumatic saddle in 1895 (#48,367) (figure 2.8g). A forerunner of the modern 'gel' saddle, this invention had some merit and may have been marketed. It reflected a concern for comfort, especially among women riders, as the safety bicycle boom took off. At that time it was widely considered unwise – for anatomical reasons – for a woman to use the kind of saddle found on men's bicycles; hence, there was considerable social pressure to come up with a more comfortable alternative.

Dr Perry Doolittle, a physician living in Toronto, was fascinated with things mechanical (later he went on to experiment with the automobile). His patents on the bicycle hub brake of 1896 (#53,808) and 1898 (#59,540) were commercially produced. The Doolittle brake required the rider to pedal backwards to engage a clutch, which shifted the brake mechanism sideways, activating a friction disk (figure 2.8h). Similar back-pedal brakes, but fitted with a lever arm attached to the chain stay, became very popular a few years later. The illustration shows the rear hub and sprocket and a cutaway of the brake mechanism. Doolittle's brake was used quite widely on Canadian bicycles.

Missing from this list is the invention that arguably ranks as the single most important bicycle-related invention originating in Canada, Thomas Willson's carbide lamp of 1896. The reason for this omission is quite simple: he never patented this invention in Canada. Born in Woodstock, Ontario, in 1860, Willson grew up in the industrial city of Hamilton, where he began experimenting with electricity and chemistry, and by 1880 he had developed early forms of arc lighting.[43] He moved to the United States in the mid-1880s to work on electrical processes for making aluminium. In 1892, while trying to develop a cheap calcium process, he serendipitiously developed the water-on-calcium-carbide process, making feasible the domestic and commercial production of acetylene gas. Calcium carbide was subsequently used for bicycle lamps, marine buoys,

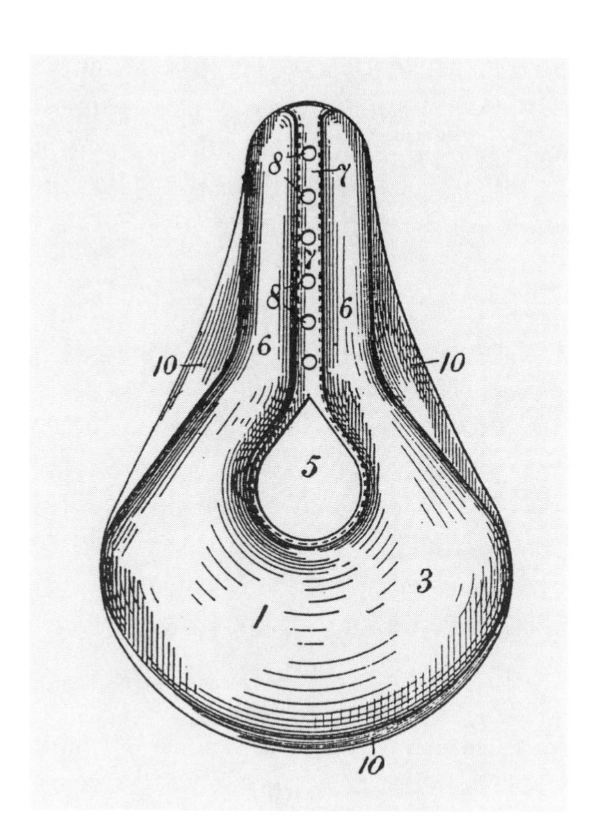

Figure 2.8. (g) Banes and Cleland's pneumatic saddle, 1895.

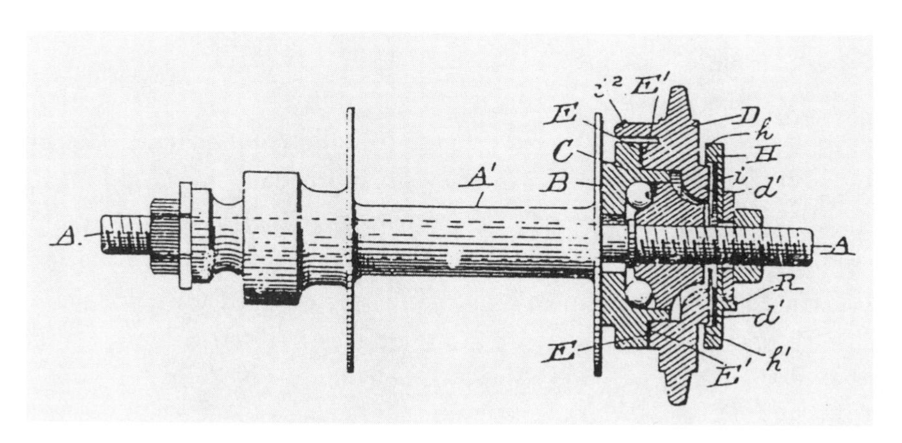

Figure 2.8. (h) Dr Doolittle's hub brake, 1896.

and headlights on early automobiles. It was also the precursor of the invention, in 1903, of the oxyacetylene welding torch. Willson returned to Canada in 1896, where he opened Canada's first carbide factory at Merritton, near St Catharines, Ontario. From this date through to the 1930s carbide lamps were widely used by bicyclists.

Accessories

During the latter years of the decade inventors were more interested in developing bicycle accessories, such as bicycle stands, locks, toe clips, footrests, shoes, umbrellas, and parcel carriers. By this time the much improved bicycle had achieved widespread social acceptance; hence, the market began to shift from the machine itself to its 'add-ons.' The various accessories patented were intended to add 'style' to the pastime, especially among the monied classes, who could best afford them. The three accessories listed below give some idea of the shift of innovative activities towards the end of the bicycle boom.

Frederick Hazard, yet another Toronto-based inventor, in 1895 patented 'a bicycling costume for women which may be worn either as a skirt, bloomers, or a divided skirt and which may be quickly changed from any one to any other' (#50,841) (figure 2.8i). The illustration shows the whole garment opened out. If it was worn as bloomers, the tapes (G) were pulled up to gather the bottom of the legs. Quite clearly, the interests of women bicyclists had risen in the consciousness of inventors by the mid-1890s, but equally, this patent reflects one of the thrusts of social modernity, the issue of 'rational dress' in women's clothing.

Boyle's Bicycle Training Machine of 1897 (#54,723) was the creation of John Boyle of Orillia, Ontario (figure 2.8j). A century later, Canadian bicyclists may be found on winter nights in their basements furiously spinning bicycles mounted on rollers, driving a fan for friction and to stop themselves overheating. Plus ça change ... Serious riders, especially during Canada's long winters, wanted some means of keeping fit and getting into shape in spring for the summer touring and racing season.

In the late Victorian era, women sought to avoid bright sunlight to keep their skin as pale as possible, so it was logical enough that someone would think of attaching a sunshade to a bicycle. Pentelow and Weston's

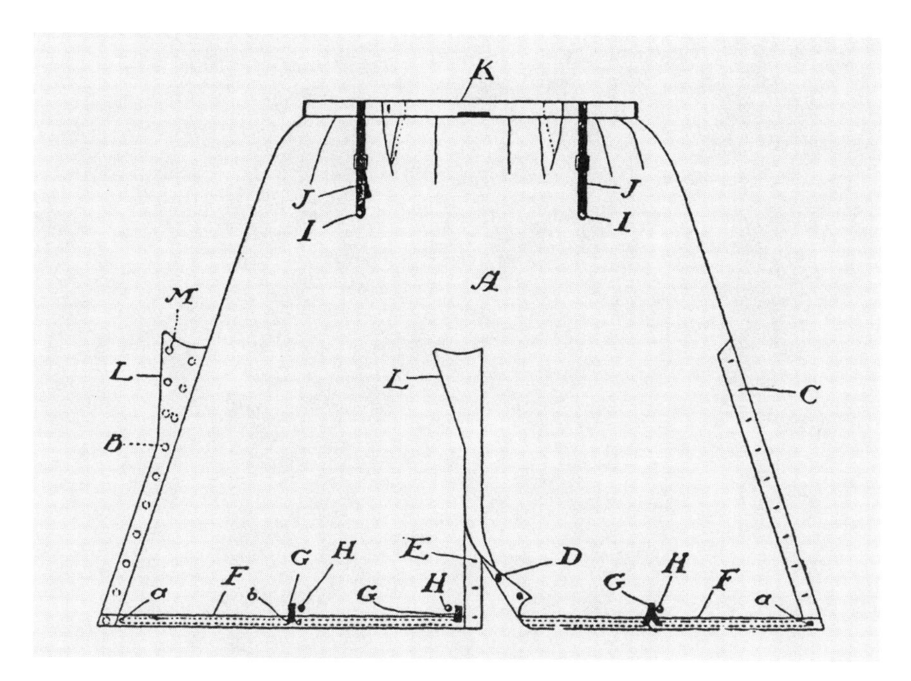

Figure 2.8. (i) Hazard's quick-change cycling skirt, 1895.

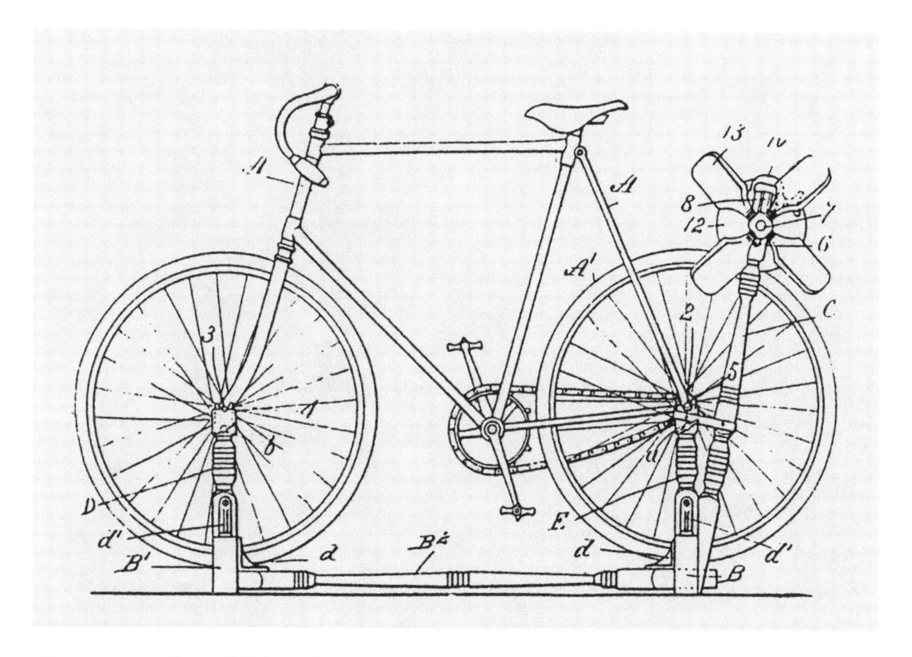

Figure 2.8. (j) Boyle's bicycle training machine, 1897.

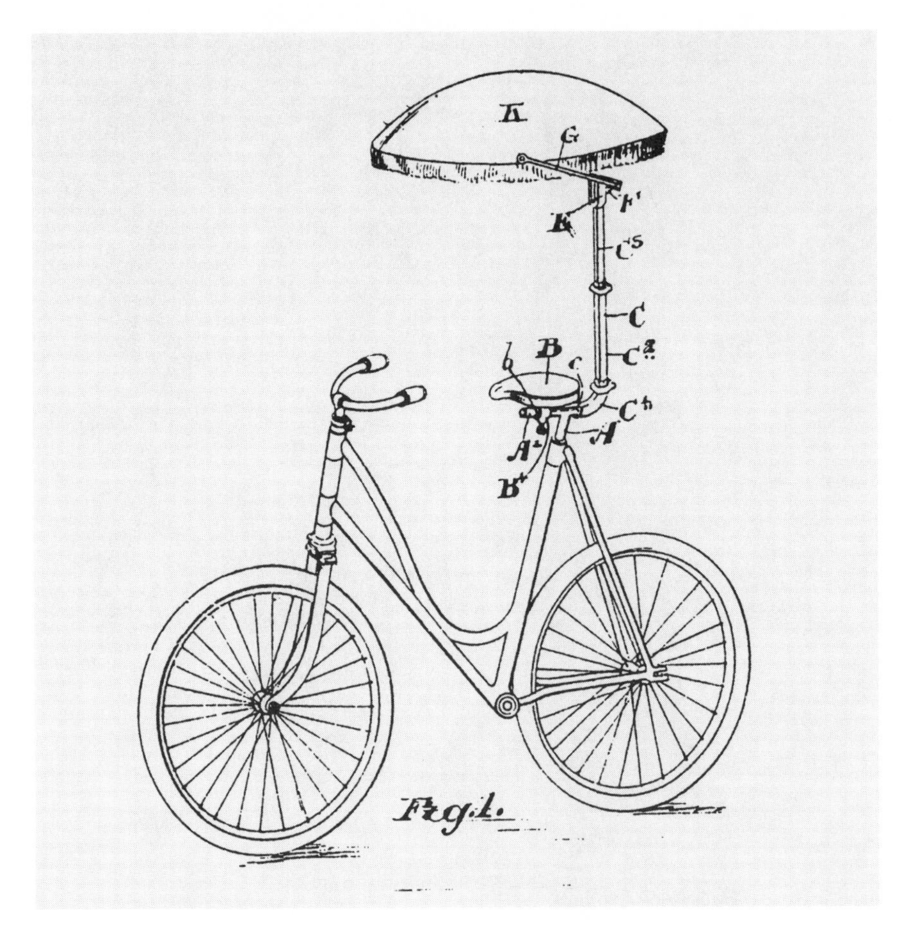

Figure 2.8. (k) Pentelow and Weston's bicycle umbrella, 1898.

Bicycle Umbrella of 1898 (#60,765) was an answer to this need (figures 2.8k and 2.9).[44] Unfortunately, even a light breeze would have sent this parasol flying, and it would be effective only when the sun was directly overhead, since there was a limited range of adjustment. Social norms, therefore, inspired this invention, just as they did with sunblock lotions a century later.

It is noteworthy that bicycle-related invention was not exclusively a male activity in the late nineteenth century. A small proportion of the bicycle patents of Canadian origin, about one in twenty, had a woman's name attached to it. Some of these inventions were for accessories, such as

Figure 2.9. An illustration of Pentelow and Weston's bicycle umbrella included in a promotional kit. Clearly an attempt was made to market this invention.

bicycling boots, an umbrella, and training wheels, but a larger number were for the bicycle itself, including Susan Fleming's gear drive, Ysobel Western's pneumatic tire, and Mary Ellen Annand's mudguard. Of particular interest is Agnes Jardine, who, with Marmaduke Matthews and Alexander Jardine, registered a series of patents for bicycle pedals, grips, spindles, and bearings.[45] Quite clearly, a small number of women were engaged in the quest for mechanical modernity and were not conforming to male stereotyping of the Victorian age.

GEOGRAPHICAL EMBEDDEDNESS

These patents have their own geography, its most striking feature being the large number of bicycle-related patents that were of non-Canadian origin. In total,55 per cent were American inventions and a further 15 per cent were British, while Canadians accounted for only 25 per cent of the total. Canada's current problem of technological dependence evidently has deep roots. Equally notable is the concentration of inventive activity in Ontario. Of a total of 275 Canadian-based patents, 221 (80 per cent) originated in Ontario. The other provinces trailed far behind, with 26 from Quebec, 6 each from New Brunswick and Nova Scotia, 1 from Prince Edward Island, 2 from the North-West Territories (which at that time included Alberta, Saskatchewan, and parts of what are now Manitoba), 2 from Manitoba, and 11 from British Columbia. It is important to remember that the geography of Canada changed rapidly during the bicycle age. In 1868 the west was still largely the domain of indigenous Canadians and fur traders, and parts of southern Ontario had yet to be opened up for settlement. By 1900 Vancouver was a thriving city, the prairies were under the plough, the Yukon had experienced a gold rush, and the transcontinental railway was a reality. It is no surprise, then, that the first bicycle patents originating in territory west of Ontario were registered only in 1896.

The contrast in the number of bicycle patents originating in Montreal and in Toronto says a great deal for the different economic roles of these two cities. At that time, Montreal was by far Canada's largest city and Toronto was a mere upstart,[46] yet Toronto accounted for 90 bicycle patents and Montreal only 14. The patent count for smaller Ontario towns, including Hamilton with 14, Brantford with 11, Ottawa with 10, London with 6, and Guelph and Woodstock with 5 each, also is revealing. At its roots, this contrast in innovative activity is connected with the differing economic cultures of Quebec and Ontario. Montreal's strength lay in commercial and financial capital linked strongly with Canada's resource industries and transportation, while its industrial capital was sectorally concentrated in textiles, clothing, footwear, and similar consumer durables. Montreal was Canada's great mercantile city and its leading port. Ontario's newer industrial and financial capital was focused more on heavy manufacturing – steel, agricultural and industrial machinery, and other engineering and technical industries. Nowhere else in Canada was a comparable manufacturing base taking shape.

WOODEN BICYCLES AND ICE VELOCIPEDES

The three waves of inventive enthusiasm for things bicycle related did not pass by without acquiring a distinctive Canadian patina. Two items stand out: ice bicycles and wooden bicycles – but then, Canada lacks neither wood nor ice and has frequently been referred to both as the 'green north' and as the 'blue north.'

Southern Canada in the late nineteenth century contained some of the finest hardwood forests on earth, forests that were being rapidly cleared to make way for agriculture. Superb timber was being cut and sold at bargain prices, while some settlers simply burned trees that woodworkers today would value highly. It is not surprising that Canadian bicycle manufacturers turned their attention to ways in which they could make use of cheap wood. At the same time, the worldwide growth of bicycle sales created a 'tube famine' as demand for steel tubing exceeded the capacity of manufacturers to supply it. Soon thereafter bicycles with wooden frames, wooden handlebars, and wooden rims appeared for sale, in addition to the more widespread use of wood for skirt guards around chains and wheels.

The wheels of the boneshaker were usually made of wood, as were the wheels of some artisan-made highwheelers, but it was in the age of the pneumatic-tire safety that the most interesting uses of wood are found. Wooden rims made of maple, ash, or beech became popular because they were light, cheap, and quite strong, though they were prone to warping if they became damp. A number of Canadian inventors patented variations on wooden rims, including the Clarksburg Wood Rim Company, which laminated strips of rawhide into wooden hoops (#62,964) in 1899. Wooden handlebars also found favour in some quarters, mainly because they were light, although they were not very strong. Most interesting are a number of wooden-frame bicycles, including the machine registered in 1897 by Charles Lavender and Thomas Fane, owners of the Comet Cycle Company of Toronto. The frame consisted of pairs of wooden stays held in place by steel brackets; only the head tube, bottom bracket, forks, and drop-outs were made of steel.[47]

The idea of riding velocipedes and other bicycles on ice seems to have dawned on northern riders quite early and is mentioned in 'Velox's' 1869 velocipede book.[48] Canadians shared this interest with other northern residents who wanted to add some spice to life during the long winter; ice-velocipede patents registered in Ottawa came from Wisconsin, Minnesota, Michigan, and New York, as well as Canada.[49] The earliest Canadian ice-velocipede patent was registered in 1869 (#3241), and thereafter such patents were submitted at frequent intervals. In total, at least thirty versions of ice-velocipedes, ice–creepers, and bicycle-sleds were recorded. The idea certainly caught the public imagination, and several illustrations appeared in Canadian magazines.

Some examples of the ice-bicycle will illustrate the enthusiasm with which this vehicle was developed. The first is the snow-velocipede of Charles Hamilton Stewart of Montreal, registered in 1869 (#3241) (figure 2.10a). Stewart describes himself as a 'manufacturer of velocipedes' and may therefore be the first Canadian bicycle manufacturer. The driving wheel on the front of his sledge might work on hard-packed snow or ice, but would likely be ineffective in deeper snow. Rather different in appearance was Walter Coburn's invention of 1891, which consisted of attachments that adapted a hard-tire safety to run on ice (figure 2.10b). These attachments could be removed in spring and the wheels reattached to return the bicycle to regular road use. Coburn lived in Toronto, which,

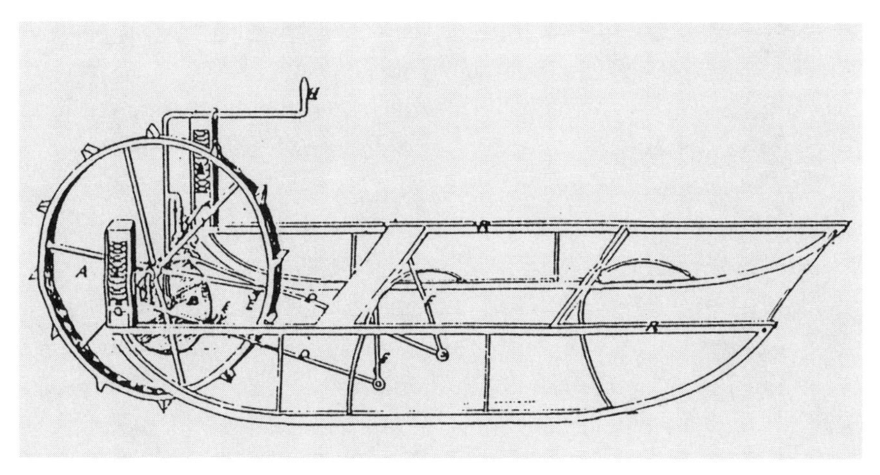

Figure 2.10. Ice-velocipede patents. (a) Stewart's snow-velocipede, 1869.

Figure 2.10. (b) Coburn's ice-bicycle, 1891.

by Canadian standards has a mild winter. Charles Casselman lived in Muskoka, however, which lies squarely in Ontario's snow belt, so the idea for his snow-velocipede of 1899 may have germinated during the long Muskoka winter. It bore a number of similarities to Coburn's machine.

The development of ice-velocipedes is far more important than might seem at first glance. These early machines were rather impractical and slow: for example, Day's ice-'creeper' (#6374) was driven by an Archimedes-type screw and could not have crawled faster than one mile an hour, flat out. As Canadian inventors sought to make their ice-velocipedes faster, they hit upon the same idea that developers of the motor bicycle did: they attached an engine, and the result was one of Canada's more dubious gifts to the modern world, the skidoo or snowmobile. The modern snowmobile, which has spawned a major Canadian manufacturing industry, is a direct descendent of these ice-bicycles.

CONCLUSION

The main conclusion to be drawn from the analysis of Canadian bicycle-related patents is that, over a period of thirty-two years, the bicycle constituted a carrier wave that affected not only bicycle technology, but also several related industries. At first, the bicycle drew on technology developed in the second long wave, but as innovation was added to innovation, the bicycle was transformed into a sophisticated vehicle that served as a platform for some of the leading industries of the third long wave, in particular, the automobile. At this level, developments in Canada mirrored those in other countries, although with a higher degree of technological dependence than in Britain, France, and the United States.

A detailed examination of the patent record suggests, however, that there were many local contingencies in the way the bicycle carrier wave found expression in Canada. Quite a number of Canadian patents, particularly those for wooden bicycles and ice-velocipedes, were embedded not only in a social and cultural sense, but also in a very explicitly historical and geographical way. The patents appeared in waves that followed the rise and fall of interest, within Canadian society, in the bicycle. The patents were highly concentrated, not only in time, but also geographically: many more patents originated in southern Ontario than in any other region, almost certainly because Ontario's manufacturing base and its workforce were better equipped to make bicycles and accessories. The environment also had a direct impact through the availability of cheap and abundant raw materials and by the impossibility of normal cycling during the long months of winter. Both Canada's vast forests and its cold winters influenced bicycle design. Note that these tendencies are not uniquely Canadian; similar tendencies were evident in the northern states of the United States, but in Canada they do seem to have been more central to the project of cycling.

Clearly, the record of bicycle patents in Canada between 1868 and 1900 is not vastly different from that of other industrialized countries during the same period, but neither is it identical. There is a series of nuances that are attributed to the fact that Canadian patents were embedded, in the broadest sense, in Canadian culture and environment. The principal conclusion is that the evolution of the bicycle was not simply a technical progression, but also an act of social construction embedded in local culture. Full development of this theme would require a systematic examination of case histories of individual patents – a project that would be difficult a century after the event – but the fragmentary evidence presented here provides evidence of the geographical embeddedness of technological innovation.

CHAPTER THREE

Modern Manufacturing: From Artisanal Production to Mass Production

Of the many currents and undercurrents that flowed with the tide of modernity, the trinity of political modernity, social modernity, and economic-cum-technical modernity had the most direct impact on modern life. For Krishan Kumar the last of these three is evidently of great importance, since it prompts him to pose the question: 'is it really possible to think of the modern world without considering that it is also industrial?'[1] The short answer to this rhetorical question is clearly intended to be in the negative. A longer answer would explore the formative link between carrier waves and industrial modernity. In this study, then, the specific links between the bicycle carrier wave and industrial progress will be examined.

The manufacture of bicycles began in small and generally unhealthy workshops, where craftsmen hammered out parts, shaved the spokes of wooden wheels, poured castings, brazed joints, and stitched leather saddles. As the bicycle subsequently evolved into a sophisticated machine, so also did the methods used to manufacture it. By 1896 large, integrated bicycle factories had been built in most of the major industrialized countries, factories that later served as models for even larger works with moving automobile assembly lines.[2] There is an almost seamless progression, therefore, from the artisan's workshop to the mass production of bicycles and cars. Unfortunately, a fragmentary and partially lost record makes it difficult to reconstruct this progression in Canada. There were plenty of Canadian artisans, but few kept an account of their operations. Moreover, in contrast to the situation in Britain, the United States, France, and Germany, only a few ordinary bicycles were produced in Canada.[3] A number of Canadian manufacturers were established in the age of the safety bicycle, but the records of the largest, the Massey-Harris Company, have been scattered and destroyed as a result of subsequent takeovers. Fortunately, enough evidence has survived that it is possible to piece together an incomplete picture. Reference will also be made to bicycle production in the United States, which at various points has had a direct impact on Canadian bicycle production.

ARTISANAL BICYCLE MAKERS

It is clear that Canadian blacksmiths, carpenters, and wheel makers wasted little time in copying the first factory-made velocipedes. For instance, the appearance in Stratford of Dr Robertson on a boneshaker in 1869 inspired two local artisans, Michael and John Goodwin, to make copies of the doctor's machine. In Toronto, John Turner, a machinist and ironware dealer, was one of several artisans who produced copies of these early velocipedes.[4] Several similar machines survive in museums and private collections, evidence that quite a number must have been made.[5]

A few of these hand-made machines were also recorded in photographs. In figure 3.1, for example, John Taylor of Fredericton, New Brunswick, is shown riding a machine he apparently built himself. The spokes and felloes appear to be iron, and both wheels are the same size.[6] The saddle sits on a spring attached to the backbone, and since only a single strut connected the front and rear wheels, it must have been extremely whippy to ride. This image suggests that there was considerable curiosity – at least in artisanal circles – in exploring the mechanical possibilities of the bicycle by making copies, and in having those replicas photographed. Production of these early velocipedes took place over a brief period; interest in them had declined by the mid-1870s.

Dr Perry Doolittle, who a decade later was to patent a hub brake of his own invention, was among the first Canadians to make an ordinary

Figure 3.1. John Taylor's velocipede, which was handmade. The design must have made it very flexible.

bicycle. Having seen drawings of one of the highwheelers displayed at the 1876 Philadelphia Centennial Exposition (World's Fair) in *Scientific Amercian*, at the tender age of fifteen he set about making a copy of it. Artisan-made highwheelers survive in the collection of the Ontario Agricultural Museum and in a number of other places, but they are generally heavy machines with inferior metallurgy and were supplanted by better, factory-made highwheelers.

MANUFACTURING SAFETY BICYCLES: THE BEGINNINGS

During the safety-bicycle era a number of Canadian bicycle manufacturers appeared. But how early? Humber suggests that Thomas Fane and

Figure 3.2. (Probably) the first illustrated advertisement for a Canadian-made bicycle. Brooks and McLean of 12 Berti Street in Toronto announced this safety bicycle in February 1891. It was a modified cross-frame safety; the crossbar could be removed for a woman rider.

Company was making his Comet bicycles by around 1890,[7] but in the very first (November 1890) issue of *Cycling* (which claimed to be the 'mirror of Canadian cycling') there are no advertisements from Fane and Company or any other Canadian bicycle manufacturer (which proves nothing, but is puzzling, since the magazine was published in Toronto). Issue no. 2 of December 1890 carries an advertisement for Brooks and McLean, manufacturers of the Planet bicycle, which may be the first advertisement for a Canadian-made bicycle, but the wording suggests that the bicycle was still in the design stage and production had not yet started. Two months later, in February 1891, the Planet advertisement in *Cycling* had increased in size to present an illustration of the bicycle (figure 3.2).

Within weeks, a series of advertisements for imported bicycles began to appear in Cycling, and the manufacturers of Gendron, Rover, Psycho, Humber, and Ecossais bicycles appointed Canadian agents. For some of the companies this was a transitional phase, to be followed by the opening of a Canadian branch factory. Hyslop, Caulfield and Company, originally agent for George Townsend's Ecossais bicycles, later manufactured its own bicycles. A year later, in April 1892, the number of pages in each issue of *Cycling* had increased substantially, and it was published every two weeks, mainly because of growth in the number of advertisements. The Wanderer Cycle Company of Toronto made a speciality of 'altering your (hard-tired) Wheel to a Cushion or Pneumatic tire,' the Goold Bicycle Company of Brantford had begun production of its Brantford range of bicycles,[8] and Gendron – one of the largest U.S. bicycle manufacturers – had set up a factory in Toronto.

By April 1897 a very different picture emerges. Cycling had become a thick magazine, and its pages were no longer numbered in order to simplify publication. Advertisements appear from the following Canadian bicycle manufacturers, with their marques (those marked with an asterisk were subsidiaries of American companies):

Cutler Brothers: Toronto	Challenge
McLean and Oakley: Toronto	Pilot
Griffiths Cycle Corp.: Toronto	Skylark
H.A. Lozier: Toronto*	Cleveland
Evans and Dodge: Windsor, Ont.*	E.&D.
James Lochrie: Toronto	Antelope
Massey-Harris: Toronto	Massey-Harris; Silver Ribbon
McBurney-Beattie: Toronto	McBurney-Beattie; Homer; Hero
McDonald and Willson: Toronto	M.&W.
New Barnes Cycle Co.: Woodstock*	New Barnes
Henderson Bicycle Co.: Goderich	Common Sense
Gendron Mfr. Co.: Toronto*	Gendron

This list is evidently incomplete: Brooks and McLean had begun to manufacture Planet cycles in Toronto in 1891; the Comet, Wanderer, G.T. Pendrith (Sun), and Griffiths Cycle companies were operating in Toronto by 1895; Goold was producing its Brantford Bicycles by 1891 or 1892, Welland Vale its Perfect Cycles in St Catharines by around 1895, and Pequegnat the Racycle in Berlin by 1897.

According to Merrill Denison, two main factors lay behind the rapid expansion in Canadian bicycle manufacturing in the mid-1890s. First, demand for bicycles had grown enormously; and second, he states that 'in 1895 the Canadian Parliament adopted its first anti-dumping legislation' by requiring 'importers to pay duty on the prices at which bicycles and parts sold in the country of manufacture.'[9] The first point is entirely true, but the second is inaccurate, even though it has been repeated in a number of subsequent studies of Canadian bicycling. In order to set the record straight, I will summarize the relevant sections of the Statutes of Canada. In broad outline, successive Conservative governments after 1867 created a protectionist regime in the name of the National Policy. Sir Wifrid Laurier's Liberal government won the 1896 election on a free-trade platform but, having won, hardly changed the Conservatives' system of protective duties.

1886: The Revised Statutes of Canada, vol. 1, chap. 33. An Act Respecting the Duties of Customs. Item 83 states: 'Buggies of all kinds, farm wagons, farm railway or freight carts, pleasure carts or gigs and similar vehicles' will pay a duty of 35 per cent ad valorem.

1888: Acts of the Parliament of the Dominion of Canada. chap. 39. An Act to amend the Act Respecting the Duties of Customs. In paragraph 1, item 11, the same description of vehicles is used as in the 1886 act (which includes bicycles), but three different duties are set:

Items costing less than $50: a duty of $10 each, and 20 per cent ad valorem;

Items costing $50 and less than $100: a duty of $15 each, and 20 per cent ad valorem;

Items costing $100 each and over: a duty of 35 per cent ad valorem.

The effect of the 1888 duties must have been to make it very expensive to import cheap bicycles and bicycle parts (on which duties evidently were also paid at these rates). For instance, the duty on an item valued at $25 would amount to 60 per cent.

1894: Acts of the Parliament of the Dominion of Canada: vol. 1, chap. 33. An Act to consolidate and amend the Acts Respecting the Duties of

Customs. In schedule A, item 380, it is stated: 'Bicycles and tricycles: thirty per cent, ad valorem.' Note that this is the first Canadian customs duty act in which bicycles are specifically mentioned as a category, and in which the rate of duty paid on bicycles and tricycles was lowered. This duty of 30 per cent was not changed in the 1897 amendment to the Duties of Customs Act introduced by Laurier's government, and it remained at 30 per cent to the end of the century. Note also that no anti-dumping legislation was tabled at this time; indeed, Jacob Viner is quite emphatic when he states: 'Canada in 1904 enacted the first general measure applicable to dumping in any of its common forms.'[10] Thus, although the rates of duty dropped slightly on expensive bicycles after 1894 and more on cheaper items and parts, they still remained at a high level in accordance with the principles of the National Policy pursued by the Conservative party in the late nineteenth century.[11] This tariff barrier was sufficiently high that it boosted the manufacture of bicycles within Canada.

What form did bicycle manufacturing take in Canada? It should not be assumed that all the emerging firms were making their own bicycles, lock, stock and barrel. In practice, there were five fairly distinct levels of operation in the industry:

Level 1. The name-badge attachers. These firms did not manufacture bicycles at all, but simply attached their own name badge to a bicycle made by another company, thereby giving the impression to buyers that they made their own bicycles.
Level 2. The frame makers. Firms in this category made bicycle frames from steel tubing to which they attached components purchased from one of the larger parts manufacturers. They were generally small operations.
Level 3. The franchises. Firms in this category were branch plants of American bicycle manufacturers. Given the import duties, there was an incentive to maximize Canadian content; some of the components of these bicycles were imported from parent factories in the United States, but others were manufactured in Canada.
Level 4. Integrated bicycle firms made frames and many of the parts themselves.

Level 5. The monopolies. These cartel-like firms emerged at the end of the bicycle boom as major firms joined forces in an attempt to stop the collapse of retail prices. In the United States, this firm was the American Bicycle Company (ABC); north of the border it was Canada Cycle & Motor (CCM).

Level 1. The Name Badge Attachers: The Normile Bicycle Company

A good example of a bicycle store that attached its own name badge to bicycles manufactured elsewhere is the W.J. Normile Bicycle and Livery Company of Napanee.[12] Normile's store was the major bicycle outlet in Napanee, Ontario. In figure 3.3 a delivery to Normile of about sixty Crescent bicycles in the late winter of 1896 or 1897 is shown. At about this time, Normile's business grew to a point where it began to order batches of bicycles with its own name badge attached.[13] Why? By selling bicycles with their own name attached, Normile sought to place itself above the category of ordinary small-town bicycle retailer. The aim was to attract more trade as a livery, cycle-repair shop, vendor of accessories, and bicycle maker, especially among the citizens of Napanee, who presumably took pride in possessing a 'local' bicycle.

Level 2. The Frame Makers: The Emperor Bicycle Company

As the bicycle craze of the 1890s gathered steam and sales of bicycles increased in leaps and bounds, growing numbers of entrepreneurs decided to manufacture bicycles. The decade was marked by a lengthy recession, which separated the second from the third Kondratieff. It created spare capacity in many plants and left manufacturers looking for new sales opportunities. There were also formidable financial barriers to surmount in setting up a fully integrated bicycle factory, although these barriers could be minimized by following the 'assembly strategy.' This strategy, which is used to this day by some of the leading customized bicycle makers, involves the manufacture of frames to which standard parts are attached.[14] In the 1890s steel tubing, lugs, drop-outs, hubs, bottom brackets, and many other accessories could easily be purchased from manufacturers, so that an energetic entrepreneur with a few hundred dollars of backing might enter into production.

Figure 3.3. Six horse-drawn sleighs loaded with crates of Crescent bicycles arrive at Normile's bicycle shop in Napanee. This delivery is being made in late winter in readiness for the spring sales boom. Hanging above the shop doorway is a highwheeler; Normile advertised themselves as agents for Gormully and Jefferies (G & J) and Dunlop tires, and for Cleveland, Crescent, and Massey-Harris bicycles.

Two such individuals were the Kalbfleisch brothers, George and Henry, of Stratford, Ontario. They were skilled machinists who had migrated south to the United States in search of work in the 1880s. Finding themselves laid off during the recession, they returned to Stratford, where in 1894 they set up a small machine shop. The brothers decided to make bicycle frames as one of their activities and created the Emperor Bicycle Company, which operated out of their machine shop. This company was essentially a frame constructor and bicycle assembler. Tubing (probably Mannesmann tubing from the United States) was made into frames to which were attached Kelly handlebars, wheels with wooden rims, and all the other necessary parts, including their Emperor head badge. In a faded photograph of the small machine shop taken in 1894 (not reproduced here) the single-storey premises are shown, with bicycles in the window and the eye-catching wheel of an ordinary bicycle on the roof.

By good fortune, some of the books recording the daily transactions of the Emperor Bicycle Company survive. The record of transactions for 1900 appears to be complete, making it possible to piece together the operation of a small bicycle-assembly firm that was also a livery (storing bicycles in winter), bicycle repairer, and machine shop doing miscellaneous work. The daybooks show that bicycle-related work accounted for about 75 per cent of the total business in 1900, peaking in spring and summer, whereas machine work rose in importance at other seasons. It is not surprising that the most frequent work they did was fixing punctures (at 35 cents per wheel), but a random selection of other repairs (with the price charged) include: cleaning a bicycle, $1; spoking and truing a wheel, $1; fitting a valve cap, 10 cents; lacing a skirt guard, 35 cents; brazing a bicycle frame, $1.50; fitting a pair of pedals, $1.75; and repairing a damaged bicycle frame, $4.

Emperor bicycles were given a serial number, as were most other bicycles. Those made in 1894 were numbered from 1; in 1895 they were numbered from 100; in 1896 they were numbered from 200, and so on. The 1900 bicycles were numbered from 600 to 626, indicating that twenty-seven Emperor bicycles were made in that year: it was clearly a small artisanal operation. It was a high quality product: the men's model, the Emperor, sold for $75 and the women's Empress for $65; in 1900 mass-produced bicycles were selling for only $30 to $50. The daybook records the sale of fourteen of these twenty-seven bicycles, plus bicycle #406, which was made in 1898 and presumably had not been sold (perhaps it was an uncommon size); it is not clear whether the other thirteen bicycles made in 1900 remained unsold at the end of the year, or whether they were sold by agents in other towns (the former may be the case, since 1900 was a year of fierce price competition).

In total, including all new and second-hand bicycles, the Kalbfleisch brothers sold thirty-three bicycles in 1900, broken down as follows:

New Emperor bicycles	15
New bicycles (other makes)	3
Second-hand bicycles	15

The monthly breakdown of sales of new and second-hand bicycles was as shown as follows:

Month	*Total*	*New*	*Second-hand*
March	2	1	1
April	5	2	3
May	14	11	3
June	7	3	4
July	3	1	2
August	1	0	1
September	1	0	1

New sales were clearly concentrated in spring, as the days lengthened and the weather improved, May being the peak month. Equally interesting, a large proportion (fourteen of eighteen) of the new sales involved trading in an old model, very like car sales today. Indeed, two of fifteen second-hand bicycles were sold with an even older trade-in. Clearly, modern patterns of planned obsolescence and upgrading were well established during the bicycle era.

The Kalbfleisch brothers continued to make bicycles until 1908, by which time they had established Stratford's first car dealership. As car sales grew, they closed down the bicycle manufacture operation. They were astute marketers; the firm made three models; the Emperor, the Empress, and the Baby Emperor. The children's model, the Baby Emperor, was launched by placing an announcement in the birth column of local Stratford newspapers.

Level 3. The Franchises: The Berlin and Racycle Manufacturing Company

Pequegnat is a name known to all Canadian antique-clock collectors. The Pequegnat family immigrated from the watch-making area of Switzerland to Canada in 1874, settling in Berlin, Ontario (renamed Kitchener during the First World War). The family was a large one, with eight sons, all of whom went into the watch-repairing and jewellery business. The eldest brother, Arthur, seems to have been the leading entrepreneurial spirit in the family; he established a jewellery business, first in Mildmay and then in Berlin. Between 1876 and 1896 Arthur also invested successfully in real estate in Berlin, at the same time helping all his brothers to establish jewellery and watch-repairing shops in towns across southern Ontario.[15] He was an amateur bicyclist, as were some of his brothers, including Paul (figure 3.4) and Joseph (who appears in figure 3.5). As the bicycle craze of the mid-1890s gathered steam, Arthur Pequegnat added a bicycle-repair and livery business at the back of his jewellery shop and then began to sell bicycles, acting as an agent for most of the leading Canadian makers.

The sequence of events whereby Arthur Pequegnat made the jump from bicycle-repairing and sales to bicycle assembly is not known, but it seems possible that it was Pequegnat who took the initiative and approached the Miami Cycle and Manufacturing Company of Middletown, Ohio, rather than the reverse. Having been granted a franchise to assemble Racycle bicycles in Canada, Pequegnat built a three-storey factory at 53–61 Frederick Street in Berlin (figure 3.6) The substantial duty Canada collected on imported bicycles was a factor in this arrangement, since making the frames in Canada minimized this tax hurdle. The Miami Cycle Company made bicycles from 1896 to 1918;[16] Pequegnat's factory made Racycle bicycles from 1897 until around 1923.[17]

Figure 3.4. Paul Pequegnat, General Manager of the Berlin and Racycle Manufacturing Company, circa 1890. He began by repairing bicycles in his jewellery shop, then retailing them before establishing the assembly operation.

Figure 3.5. The Stratford Bicycle Club meets in Brantford, Ontario, in 1888. Interesting bicycles in this picture include a loop frame tricycle on the extreme left, and a Kangaroo (the second upright bicycle from the right). The fourth from the right is one of the first hard-tire cross-frame safety bicycles seen in Canada. The smaller bicycle lying on the ground (second from the left) is an American Light Safety. Joseph Pequegnat is standing, centre row, fifth from the right.

Figure 3.6. The Berlin and Racycle Manufacturing Company's factory in Kitchener, which later became a clock factory. The giant crank on the post outside the works reminded passers-by of the distinctive crank fitted to Racycle bicycles.

The Canadian franchise began operations under the name 'The Berlin and Racycle Manufacturing Company.' Pequegnat chose well in that the Racycle was a high-quality bicycle with a unique feature, which, it was claimed, saved one-fourth of a rider's strength: the bearing on the crank hangar was placed outside the crank attachment, so that pressure on the chain sprocket was spread more evenly to both bearings.[18] This claim is mechanically suspect, but the bicycles were well made, with a range of men's, ladies, and children's models (ten models are listed in their 1906 catalogue). Miami itself did not make the entire bicycle – it used Goodrich tires, Kelly handlebars, Standard pedals and spokes, and gave purchasers the choice of a Person, Troxel, or Wolverine saddle and several different coaster brakes. Perhaps the best known of the Racycle models was the Pacemaker, which had a very large (15-inch diameter) chain wheel, which looked more impressive than it was (it resembled a track bike but, because of its larger rear sprocket, was, in practice, somewhat lower geared).

The Berlin and Racycle Manufacturing Company produced solid and reliable bicycles, which sold quite well in Berlin, Waterloo, and the surrounding area. Among its advocates were groups that valued reliability; for example, the local police department in Kitchener used Racycles for over thirty years.[19] The bicycle achieved some recognition overseas, and exports accounted for a significant portion of the company's sales. In reality, however, the arrangement was an early version of Canada's branch plant economy. Protection from the dumping of cheap American bicycles and parts helped Canadian bicycle manufacturers to a degree, but some American manufacturers, including the Miami Cycle Company, responded to Canadian tariffs by setting up franchises and branch plants inside Canada.

A Digression: Bicycle Manufacture in the United States

Before integrated Canadian bicycle manufacturers are discussed, a digression will be made, because the activities of the Pope Manufacturing Company, which was the largest bicycle manufacturer in the United States (and the world) during most of this period, are relevant to what happened in Canada. As with so many other economic affairs, the Canadian bicycle industry cannot be fully understood without looking at the larger continental setting; Canadian bicycle companies copied, competed with, and purchased many of their parts from American firms. Indeed, the Canadian firm to be discussed, Massey-Harris, had close business connections with Pope's firm.

The dominant position that British bicycles had occupied in the Canadian market during the early part of the bicycle era was increasingly challenged by American bicycle manufacturers in the late 1880s – indeed, by the 1890s American companies held the largest share of the Canadian market. This conquest is attributable to the modernization of the American bicycle industry, which increased the productivity of its workers and produced a stream of innovations both in the design of bicycles, and in their methods of manufacture and sales. In due course, Canadian bicycle manufacturers followed some of the pathways taken by the larger U.S. companies.

The 1876 Centennial Exposition in Philadelphia seems, with hindsight, to have been a seminal event in American bicycling history. The display there, and presumably on the roads around the exhibition, of a variety of British high bicycles seems to have been one of those timely events that capture the public imagination. Among those who saw commercial possibilities in the highwheeler was Colonel Albert Pope, a relatively small-scale New England manufacturer and venture capitalist, who decided to market his own highwheeler, largely by copying one of the displayed bicycles (the Duplex Excelsior made by Bayliss, Thomas of Coventry, England). In 1877 Pope subcontracted the manufacture of a batch of fifty bicycles to the Weed Sewing Machine Company of Hartford, Connecticut, which was running well below capacity in a depressed sewing-machine market and looking for new products to keep its workers busy. Pope's 'Columbia' bicycles sold well, and year by year Pope improved his bicycles and expanded his range of models. Pope's success is attributable to his contribution to industrial modernity on a number of fronts, five in particular.[20]

1. Pope progressively modernized the labour process by extending the division of labour in his factories and, wherever feasible, mechanizing each of the simplified steps. By 1894 there were 840 parts in a Columbia men's bicycle and nearly 1,000 parts in a woman's bicycle, and over 500 inspections were made during their assembly. In the late 1890s, as the bicycle industry went from boom to bust, Pope cut his prices, mechanized further, and trimmed his workforce wherever possible, practising a form of lean production decades before the concept entered the vocabulary of management schools.
2. Pope also advanced the system of vertically integrated production. Initially, being strapped for cash and unsure of how well his bicycles would sell, he subcontracted production to the Weed Sewing Machine Company, which made bicycles for Pope in batches. By 1890 Pope had purchased Weed outright; he then proceeded to buy out the Hartford

Rubber Works (to make tires). Next, he built a seamless tubing factory, added a separate assembly line for his lower-priced Hartford range of bicycles, and in 1894 built a head-office building adjoining the factories to create the world's largest vertically integrated bicycle complex.

This integration went further as the demand for bicycles declined after 1897. Forty-two of the leading U.S. bicycle manufacturers agreed to form a 'bicycle trust' or cartel, named the American Bicycle Company (ABC), which was supposed to stop the collapse of bicycle prices. As will be shown presently, the decision in 1898 by ABC to build a bicycle factory in Hamilton, Ontario, precipitated a merger wave in the Canadian bicycle industry. ABC went into receivership in 1902 and was resurrected as the American Cycle Manufacturing Company.[21]

3. As noted earlier, the bicycle created a minor carrier wave in the latter part of the nineteenth century. Pope seems to have anticipated this trend: he was innovative in bicycle design and kept a watchful eye on developments introduced by other bicycle makers in the United States and Europe. He also played a wily game with patents: he was careful to register inventions made by his own company and would purchase useful patents from others if he felt they were commercially applicable.

Although David Hounshell, in his 1984 book *From the American System to Mass Production, 1800–1932*, recognizes Pope's contribution to quality control and testing, he sees Pope's overall approach to manufacture as rather conservative 'in the Yankee armoury tradition.'[22] Columbia bicycle parts were made in the same way that New England arms makers made gun and rifle parts – by milling and machining away metal until it was exactly the right size. This was a slow and work-intensive process that wasted metal. The approach developed by carriage and agricultural machine makers in the American midwest was more innovative: copying the methods of these firms in 1896, the Western Wheel Company of Chicago, maker of Crescent bicycles, adopted stamping and pressing technology that greatly accelerated the production of parts and lowered their cost. In consequence, sales of Crescent bicycles briefly overtook those of Columbia.

In other respects, Pope was innovative. For instance, his factory in Hartford was the first to use an Otis elevator for industrial purposes.

More important, Edison built for Pope the first electrified continuous-production assembly line, which was visited and admired by Henry Ford.[23]

4. Womack, Jones, and Roos stress that the key to mass production was not the moving assembly line, but the interchangeability of parts.[24] Columbia bicycle parts fitted exactly. Yankee machining technology, coupled with numerous inspections, guaranteed this precision. Pope took advantage of interchangeability by using the same part in as many different models as possible.
5. Pope's greatest contribution to mass production was to recognize that it required mass consumption. Without a large market, it was pointless to make large numbers of bicycles. To this end, Pope put great store on advertising Columbia bicycles both commercially and through other forms of promotion that were intended to persuade the public that the bicycle provided one of the most accessible pathways to modernity. For instance, he provided a security of $60,000 for the start-up costs of the *Wheelman* magazine; he argued vociferously for the right of bicyclists to use public parks (and contributed thousands of dollars to litigation resulting from contrived infractions of ordnances forbidding bicycling in Central Park, New York); and he invested huge amounts of energy in promoting the Good Roads movement.

Level 4. The Integrated Bicycle Manufacturers: Massey-Harris Cycles

It seems likely that Massey-Harris was the largest producer of bicycles in Canada in the late 1890s, followed by H.A. Lozier, the maker of Cleveland bicycles. Both firms were key players in the CCM merger.

Massey-Harris for some years had been Canada's leading agricultural machinery manufacturer, when in 1895 a decision was made to diversify into bicycles.[25] Two representatives of the company toured the factories of many of the leading U.S. bicycle manufacturers, closely examining model design and methods of production. They recommended that Massey-Harris secure the Canadian rights to the patents of Columbia bicycles owned by the Pope Manufacturing Company, so that Massey-Harris's early models would bear a family resemblance to the Columbia. Massey-Harris located its bicycle factory in Toronto on King Street West (figure 3.7), although in practice only part of these premises was used to make bicycles.

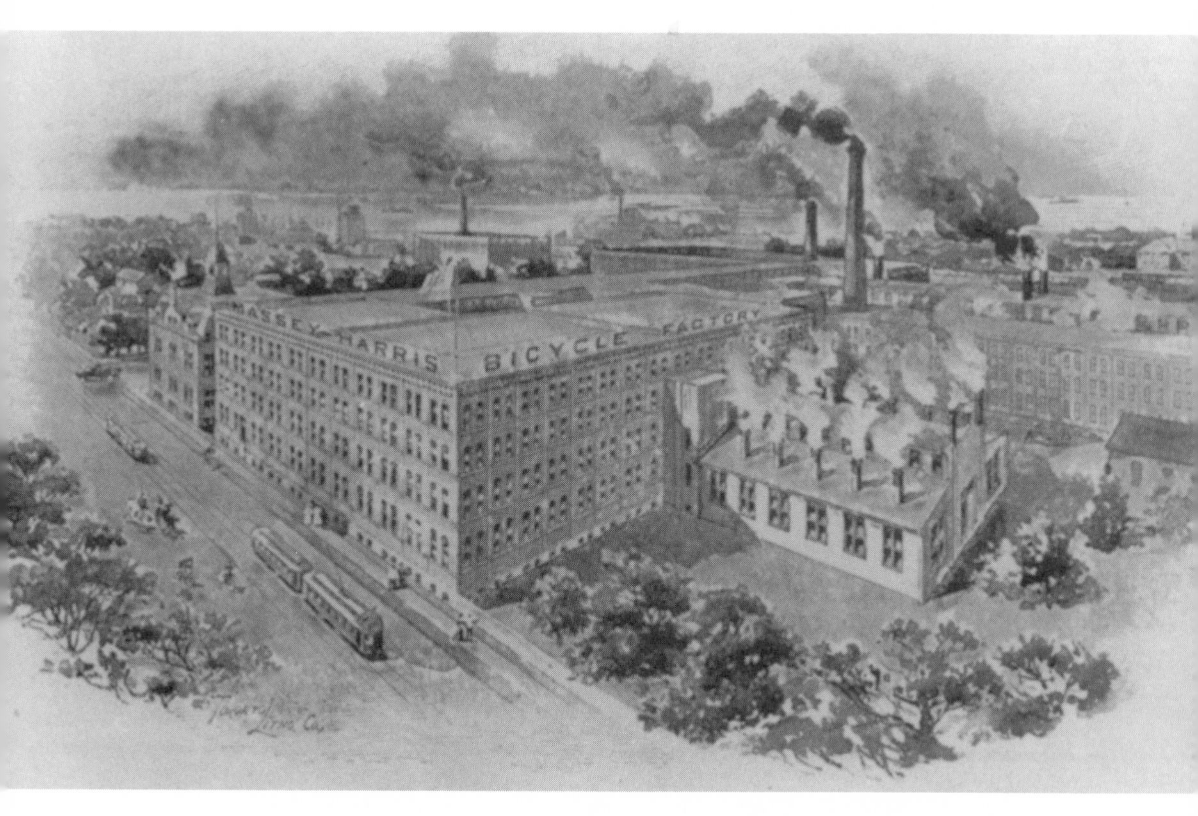

Figure 3.7. The Massey-Harris bicycle factory in Toronto. It was normal for nineteenth-century manufacturers to magnify the scale of factories in their brochures. Here the illustrator depicts streetcars, bicycles, and trees at about half their real size, while pedestrians on the sidewalk appear like ants.

Being Canada's largest agricultural machinery manufacturer, Massey-Harris had the resources to launch its bicycle manufacturing operations on a large scale. In 1896, the first year of production, Massey-Harris produced two models, the Gentlemen's Model 1 and the Ladies' Model A, both priced at $85. A year later they were superseded by the Model 2 and Model B, respectively, now with the option of a brake and a choice of colours – maroon or black. By 1898 the Model 3 and Model C were offered, in several different frame sizes and with a choice of saddles and handlebars. In addition, two new models were offered, the Light Roadster (Model 4) and the Road Racer (Model 5). In 1899 an addition was made to the bicycle factory on King Street; the Model D now was equipped

Figure 3.8. Drop hammers in the Massey-Harris smith shop.

Figure 3.9. The machine shop at the Massey-Harris factory. Rows of belt-driven machines mill away metal until hubs, brake parts, axles, cranks, and other parts meet the 'specs.' The specifications of parts were set on these automatic milling and machining tools, so that the machinist became a machine minder.

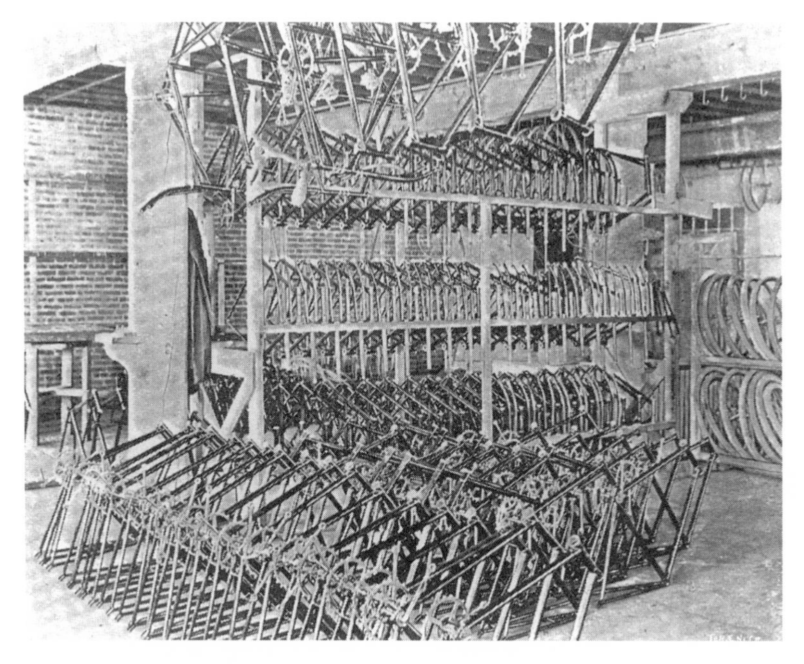

Figure 3.10. The Massey-Harris frame shop, with frames ready for final assembly.

Figure 3.11. The assembly shop at the Massey-Harris factory in Toronto.

with a skirt guard, and the three men's models were superseded by Models 6, 7, and the '7 Special' (a light racing bike). Note the strategy of regularly introducing new models, thereby encouraging customers to discard their old models.

A number of photographs survive that illustrate manufacturing operations in the Massey-Harris works, and help to reconstruct the steps in the production of their bicycles. In figure 3.8 drop hammers in the smith shop are shown, hammering out parts to their approximate dimension. The parts were then taken to the machine shop, where they were shaped to exact specifications. The most advanced machining, done by the 'automatics' shown in figure 3.9, represented a significant step in transferring skills from the machine operator to the machine.

The frames of Massey-Harris bicycles were made with 'Mannesmann Welsh Special Cold Drawn Seamless Tubing,' imported from either the Pope-Mannesmann tubing factory in Hartford, Connecticut, or from Britain (figure 3.10). The set-up of assembly operations at the time of the merger into CCM in 1899 is shown in figure 3.11. The scale of operations pales in comparison with the large assembly lines at Pope's factories located in Hartford, but there is, nevertheless, a clear division of labour.

Level 5. The Monopolies: The Canada Cycle & Motor Company (CCM)

During the mid-1890s, at the height of Canada's bicycle boom, competition among the many small bicycle manufacturers was fierce. Thereafter, these small manufacturers began to consolidate, following a pattern very similar to that adopted south of the border.

There were only seventeen establishments manufacturing bicycles in the United States in 1890. Five years later this figure had risen to three hundred factories, an astonishing seventeen-fold increase, despite the recession of 1893–94. The American fascination with the bicycle peaked in 1896 and 1897. About 1.2 million bicycles were manufactured in 1896, rising to close to 2 million in the following year as prices began to drop.[26] Thereafter the bloom was off, and sales fell to around 1.1 million in 1899 and 1900, creating considerable surplus capacity and leading to idle plant and machinery. Faced with a palpable fading of interest among the ranks of society's trendsetters and the pressing need to arrest the decline in bicycle retail prices, American bicycle manufacturers began to consolidate. By far the most important of these merger moves was the decision made by forty-two American cycle manufacturers in 1898 to merge so as to form the American Bicycle Company, ABC.

At this time the Canadian bicycle industry was still fragmented, but the decision made by ABC in 1898 to build a bicycle factory in Hamilton, in Ontario's industrial heartland (and inside Canada's tariff wall), sent a wave of panic through the Canadian industry. Would Canadian cycle manufacturers be able to compete with this aspiring American multinational? Would ABC underprice Canadian manufacturers? Fearing the worst, a group of leading Canadian bicycle manufacturers decided to follow the American model and consolidate. Why did ABC decide to locate in Canada? Denison's explanation – that it was a response to the anti-dumping legislation of 1895 of Sir Mackenzie Bowell's government – is inaccurate.[27] Quite simply, the import duties were high enough to make it worthwhile for several American bicycle manufacturers to open branches in Canada, including Gendron in Toronto in 1892, H.A. Lozier & Company (maker of the Cleveland bicycle) in Toronto Junction in 1895, Evans and Dodge in Windsor in around 1897, and ABC in Hamilton in 1898. Racycle found a Canadian partner to assemble its bicycles in 1897.

Canada Cycle & Motor Company Limited was formed in 1899, with Walter Massey as president, George A. Cox (president of the Canadian Bank of Commerce) as vice-president, plus Sir Joseph Flavelle (like Massey, one of Canada's leading Methodist entrepreneurs), E.L. Goold (of the Goold Bicycle Company), E.R Thomas of the H.A. Lozier & Company, and two others as directors. Its mandate was to merge the five leading Canadian bicycle firms. The firm's total capitalization was $6 million, of which half was common stock and half preference shares. For the preference stock, $500,000 had been subscribed by the directors, $500,000 was retained in the treasury of the company for future operations, and the balance was offered for public subscription; $3 million was issued as common stock and was oversubscribed.[28] Given that the five companies collectively had made profits of $191,000 in 1896–7, $305,000 in 1897–8, and $330,000 in 1898–9, the prospects looked good.

CCM – the Canada Cycle & Motor Company Limited – was formed in September 1899.[29] The five companies participating in the initial merger were: The Massey-Harris Manufacturing Company Limited of Toronto, proprietors of the Silver Ribbon trade name; H.A. Lozier & Company of Toronto Junction, makers of Cleveland bicycles; the Welland Vale Manufacturing Company of St Catharines, who manufactured the Perfect bicycle; the Goold Bicycle Company Limited of Brantford, makers of Red Bird bicycles; and the Gendron Manufacturing Company Limited of Toronto.

As fate would have it, ABC ran into financial trouble in the early years of the new century, owing to declining sales of bicycles in the United States. The total value of American bicycle production declined from $31.9 million in 1900 to $5.2 million in 1904, the number of factories shrank from 312 to 101, and there were massive lay-offs as the industry's workforce fell from 17,525 to 3,319. ABC was a victim of this catastrophic decline; it sold its Canadian subsidiary located in Hamilton, known as the National Cycle and Automobile Company, to CCM in 1902. At the same time CCM also acquired the Evans and Dodge (E.&D.) bicycle factory in Windsor, and the rights to twenty-two registered bicycle brand names held by E.&D.

CCM also ran into serious financial difficulties at the turn of the century, although the Canadian bicycle boom peaked slightly later than it did in the United States, and the collapse of the industry was not quite so catastrophic. CCM sales over this period were very approximately as follows[30] (with total U.S. sales estimates by Perry in parentheses):[31]

Year	*CCM Sales*	*U.S. Sales*
1898	38,500 (Massey-Harris only)	n.a.
1899	60,000 (CCM)	(1,182,691)
1900	44,000 (CCM)	(1,182,850)
1901	15,000 (CCM)	n.a.
1902	20,000 (CCM)	n.a.
1903	25,000 (CCM)	n.a.
1904	50,000 (CCM)	(250,487)

CCM survived the crunch by restructuring and shrinking and with financial help from Massey-Harris and its directors. The sales promotion teams of the five companies were merged into one team to avoid duplication, and there were many lay-offs. Leading shareholders advanced security to keep the company afloat. Production was rationalized first by moving the manufacture of Massey-Harris and Gendron bicycles to the factories in Toronto Junction, St Catharines and Brantford. When the slump definitely set in, production was concentrated at the former Lozier plant in Toronto Junction. In 1917 CCM moved to a large, new factory in Weston, at a time when the bicycle was making a comeback – but that development lies outside our period. CCM was also involved in the automobile trade: for a short while, it was the Canadian agent for the American Winton car, the Waverley electric car, and several others. Then, in 1905 CCM began to make its own car – the Russell – thereby following the transition that many other bicycle makers had made, most notably Henry Ford.[32]

CCM sold bicycles not only in Canada, where it partially reclaimed the market from American and British manufacturers, but also in foreign markets. Karl Creelman, whose epic ride is described in chapter 7, below, rode and promoted the Red Bird bicycle in Australia, where a network of Brantford agents was well established (figure 3.12). Jim Fitzpatrick notes that Canadian-made bicycles were popular in Australia at this time.[33] The British Empire was the particular target of these sales efforts, and a series of advertisements that appeared in Cycling in the spring of 1901 played heavily on this theme. In figure 3.13 the good ship Canadian Bicycles (so identified in another advertisement) is shown sailing into harbour with the slogan 'Good Things from the Colonies.' A similar ad plays on this colonial theme by including a Union Jack with the phrase 'Made under the Flag,' placed below it. It made commercial sense, and often still does, to connect products to patriotism. The advertisement printed on 18 May 1901 (figure 3.14) reminds the reader of another Canadian invention – 'Carbide' Willson's acetylene lamp, which is placed on the woman's bicycle.

The South African War of 1899–1902 gave CCM an unusual opportunity to boost sales of its bicycles, playing simultaneously to several markets. Under the headline 'Embarking for the Front,' the text shown in figure 3.15 states that the cycle corps of the Fifth Contingent Queensland

Figure 3.12. This allegorical image from the Goold Bicycle Company's 'Red Bird' catalogue of 1898 shows Miss Canada (with a beaver beside her) passing a bicycle to Miss Australia, who is accompanied by a kangaroo. It follows another allegory showing the same Miss Canada shaking hands with Britannia over the caption, 'CANADA: A worthy daughter of a worthy mother.' There was a very conscious play by Canadian manufacturers on imperial trade relationships during the bicycle boom.

Figure 3.13. This 1901 advertisement shows SS *Canadian Bicycles* sailing in to harbour loaded with CCM bicycles. Made under the flag? The Maple Leaf? Not at all, the Union Jack! CCM, a 'colonial enterprise,' played on that theme to promote sales of its bicycles.

Figure 3.14. Modernity from the colonies! A Canadian cycle, lit by the latest carbide lamp, rolls into the picture beside an electric streetcar. By 1901, women's skirts were significantly shorter than in the mid-1890s.

Figure 3.15. In support of the British war in South Africa, Australian troops equipped with Canadian bicycles are shown lined up beside the ship that will transport them to Cape Town, where they took part in the first war in which bicycles played a significant role. This advertisement makes a strong pitch to British, Australian, and Canadian markets to foster intercolonial trade.

Imperial Bushmen had been equipped with Massey-Harris bicycles after strict examination and trial of several different machines. The ink drawing that accompanies the advertisement shows rows of soldiers standing to attention beside their sturdy Massey-Harris bicycles, waiting to board a ship bound for Cape Town. The sales implications were wonderful: clearly, other cycle corps should also be equipped with this superb machine; all Australians should follow the Imperial Bushmen's lead and buy Massey-Harris bicycles; the British should show appreciation to their colonial partner in the war by purchasing quantities of CCM bicycles; and Massey-Harris should be congratulated for its noble contribution to the war effort (the first war, incidentally, in which bicycles played a significant role).

CONCLUSION

In the United States, Britain, France, and Canada industrial modernity found expression in new products and new methods of manufacture. Some of the more important innovations stimulated demand on a sufficiently large scale that changes were required to the organization of production. In the space of thirty years, dramatic changes took place in the way bicycles were made. The scale of operations grew from small workshops employing a few craftsmen to large vertically integrated factories. The mode of production, which began as an artisanal form, finished, at least in the case of the largest American manufacturers, as a prototypical system of mass production. The division of labour was extended through the workforce. Work became more regimented and oppressive, but productivity increased. The workforce began to organize. Manufacturing technology evolved from hand-working techniques to semi-automated processes with linear assembly lines in electrified factories. Industrial organization began with highly competitive small producers and concluded with large firms like CCM.

Industrial modernity advanced on many fronts, but especial importance is attached to the bicycle industry because its impact reverberated through other sectors of the economy. As industrial wages rose and bicycle prices came down, so the machine became affordable to an increasingly large public. By the end of the century, photographs show a far broader spectrum of society riding bicycles than had been the case even five years earlier. The noticeable decline in the passion of the social elite for bicycling after 1897 was probably more a function of working people joining the ranks of bicyclists than anything else. By 1900 mass production methods made it possible to make a 'lower-grade' bicycle for about one-third of the price of an 1895 model. Industrial progress destroyed the exclusivity of bicycling as new forms of manufacturing made cheaper bicycles available to a mass market. The social elite responded by moving on in search of new activities on the cutting edge of fashion and technology.

CHAPTER FOUR

Bells and Whistles: The Bicycle Accessory Industry

> **Bells and whistles**: noun, informal – additional features, typically superfluous, but superficially attractive
>
> John Ayto, *The Longman Register of New Words*

The term 'bells and whistles' came into common use as recently as the mid-1980s, when journalists began to employ it to describe the new generation of microcomputers.[1] In that highly competitive market, computer manufacturers started adding extras to woo the potential buyer. What was the origin of this term? In the 1991 edition of *The Oxford Dictionary of New Words* the opinion is offered that the term was an allusion to fairground organs with their multiplicity of bells and whistles, but it could equally be a reference to bicyclists of the 1880s, who also made extensive use of bells and whistles. In figure 4.1 a wheelman is shown in his club uniform standing beside his high bicycle. On the front of his tunic hangs a bicycle whistle, while a bell is attached to the front wheel of his ordinary.

Bells and whistles, and all the other accessories that bicyclists purchased, were of immense economic significance and added greatly to the size of the bicycle carrier wave. Smith reports an estimate made in May 1896 that to that date Americans had spent $200 million on bicycle sundries and repairs and $300 million on bicycles themselves (including

Figure 4.1. The year is 1885, and Horace Joyce is dressed in the uniform of the Montreal Bicycle Club to pose in William Notman's photographic studio. A Hill and Tolman bell is attached to the bicycle's front forks, while a whistle hangs from his tunic. These bells worked on a wind-up system, so that every ten to twenty seconds they would ring once. The whistles were often used by captains on club rides to give orders to riders, especially in formation riding. It is a moot point whether bicyclists used these warning devices more to draw attention to themselves than to warn pedestrians and cab drivers of their presence.

accessories on new machines).[2] If we conservatively estimate that one-sixth of the latter (i.e., $50 million) was accounted for by accessories, then we arrive at a fifty-fifty split. Roughly half of all bicycle-related expenditures were on accessories, sundries, and repairs. The sales of bells and whistles – tool bags, bicycle clothing, maps, lamps, and locks – and the purchase of services such as riding lessons, accommodation, club memberships, and bicycle insurance doubled the economic impact of the bicycle. These expenditures gave rise to a ditty that appeared in California's *Riverside Daily Enterprise* on 25 August 1896:

Hey diddle diddle
The bicycle riddle,
The strangest part of the deal;
Just keep your accounts –
Add up the amounts;
The 'sundries' cost more than the wheel.

The bicycle was not the first consumer good to be accompanied by a considerable sale of sundries (to use the term current in the bicycle era). Among other goods that had stimulated a substantial trade in such items were sewing machines, children's toys, carriages, and microscopes. However, the sales of add-ons were of even greater significance in the bicycle industry; indeed, a case can be made that they represented a key transition, heralding a new phase of modern consumption that was to blossom in the late twentieth century into what could be labelled 'Barbie Doll' industries, where consumers are initially encouraged to purchase a product, often at a subsidized price, because the main profits are made subsequently on the sales of innumerable add-ons.[3]

Put succinctly, the argument is as follows. The bicycle age constituted a minor carrier wave, which advanced the division of labour and production technology in large, integrated, bicycle-manufacturing and assembly complexes. Since the initial purchase of a bicycle represented only half of the total outlay, with the rest being accounted for by the purchase of accessories (defined to include all related purchases), the bicycle must have influenced consumption in diverse ways. Although consumers were not bombarded with advertising to the same degree as they are today, the industry nevertheless managed to promote bicycles, accessories, and sundry services in various sophisticated ways.

One of the main uses of these accessories was to attract the attention of the crowd on the street or in the park. Having purchased what was one of the most visible markers of modern consumption, the bicyclists of the 1880s and 1890s were determined that the public should take notice of them. Having created a carrier wave, the producers of bicycles and ephemera also were eager to boost that wave by selling the consumer all manner of accessories and updates. Consumer industries of today, including, for instance, the manufacture of children's toys and of golfing and skiing equipment, are replicating what had happened on a similar scale during the bicycle era a century earlier.

BICYCLE ACCESSORIES: THE HIGHWHEELER PHASE

A distinction can usefully be drawn between the more limited range of sundries of the highwheeler phase, and the proliferation of accessories during the era of the safety bicycle.[4] Four main factors account for this distinction. First, the ordinary was very rarely ridden by women, whereas in the safety era a range of different accessories was marketed for men and for women. Second, the smaller volume of sales of the high bicycle meant that it was simply not economic to produce certain accessories that became economic later because of the much bigger volume of sales of safety bicycles. Also, the growing social cachet of the safety bicycle in the 1890s led to an increasing interest in acquiring distinctive bicycle accoutrements. Finally, changes in technology and in the rules and regulations of bicycling produced new possibilities and new demands in the safety era.

Two different types of evidence will be used to demonstrate the most common accessories of the highwheeler era: first, the contents of the 1887 catalogue of one of Toronto's leading bicycle distributors, Charles Robinson and Company will be assessed; and second, a selection of photographs from this era will be used to illustrate particular items.

In 1887 Charles Robinson of 22 Church Street in Toronto was the sole Canadian agent for Rudge and Company, at that time Britain's largest bicycle manufacturer. On the first four pages of the catalogue, Robinson advertised nine different machines: the Rudge Nos 1, 2, and 3 ordinaries (priced at $115, $85, and $60), the Kangaroo-like Rudge Safety ($115), an early cross-frame safety called the Rudge Bicyclette ($115), the Boy's Ideal ordinary ($32–$42), the Boy's Velocipede ($7–$10), the Rudge Humber Tandem Tricycle ($200), and the Rudge Rotary Tandem tricycle ($175). The next four pages (i.e., fully half the catalogue) is a 'condensed' list of accessories and sundries containing ninety-seven separate items. The allocation of over half the catalogue to such accoutrements lends credence to the argument that they were a central feature of bicycle consumption. Among the more interesting were a range of bells (also called alarms), an anti-header device, various lamps for night riding, luggage carriers, spoke tighteners, whistles, and rust preventatives. Some of the accessories were not cheap: the nickelled Lucas 'King of the Road' hub lamp cost $6, the Butcher improved cyclometer $11, replacement rubber tires that were cemented to the rims $6, a good bugle $10, and saddles from $3 to $7. Replacement front bearings, which wore out if a rider did a high mileage, cost a substantial $10 per pair. In figure 4.2 a selection of items in Robinson's list are illustrated; both the Cyclist's Wallet ($1) and the King's Own Tool Bag ($3) were attached to the saddle, the tool bag costing more because it contained a pair of pliers, a padlock and chain, a canteen, and wire. The clamshell tool bag made by 'Don Tool' also attached to the saddle. Bicycle shoes and long socks were important items of clothing. Club uniforms were supplied by tailors. Note the design of the shoe, its long lacing, narrow arch with comparison to modern styles, and leather cleats on the sole indicating that high-bicyclists pedalled on the balls of their feet.[5]

When cyclists visited photographers they often took with them not only their machines, but also a collection of accessories. In figure 4.3 a group from the Ottawa Bicycle Club display a range of sundries. The cyclists are wearing bicycling shoes and socks and either a pith helmet or a suitably fashionable hat. Hanging from their tunics are other items – a club badge and chains that may have a watch or whistle attached. The bicycle on the left has a Lucas hub lamp hanging from the front axle. The wheelman kneeling in front has his hand on a canvas cycling bag, while the person behind him has a similar bag suspended

CYCLING SUNDRIES.

Bicycle Bags.

THE CYCLIST'S WALLET.

A commodious and well made bag, manufactured by Lamplugh and Brown.

Price.............................$1 00.

The King's Own Tool Bag.

Containing pliers, padlock and chain, canteen and wire.

Price.............................$3 00.

The Don Tool.

A favourite bag with cyclists.

Price.............................$1 50.

Oil Cans.

Lucas' King's Own, with stop cock.

Price.............................40cts.
Ordinary cans.....................25cts.

The Perfection Bicycle Shoe

This is the best Bicycle Shoe manufactured, being made of the best grained calf skin. It is laced nearly to the toe, thus being easy on the foot; it has leather pleats on the sole so arranged as to catch the pedals and thus preventing slipping; it is made in half sizes, that is, No. 6, No. 6½, No. 7, No. 7½, etc., and in two widths. It is much superior to the English or American shoe and the price is only $3 50.

Bicycle Hosiery.

28 in. long, in gray, blue or any color, $1.25 and $1.50

Locks and Chains.

Yale bronze lock, chain and key. Price......$1 25.
Steel locks.......................25 and 50cts.

Harwood's Adjustable Step.

HARWOOD'S SAFETY BICYCLE STEP

Price.............................$1 25.

SEND FOR PRICE LIST OF

LACROSSE,
BASEBALL,
CRICKET,
FOOTBALL,
LAWN TENNIS,
AND OTHER SPORTING GOODS.

Figure 4.2. (opposite) Some of the highwheeler sundries advertised by Charles Robinson & Co. of Toronto in its 1887 catalogue. The step was attached to the backbone above the rear wheel to help riders mount; some riders added a second step to make mounting even easier. Long socks and shoes were important to prevent trousers from catching in the spokes. Highwheel riders were enthusiastic oilers of their machines. The locks suggest that thieves were quick to discover a new market.

Figure 4.3. (above) A studio portrait (circa 1886) by Pittaway and Jarvis of Ottawa. Three members of the Ottawa Bicycle Club pose in tailor-made club uniforms, bicycle shoes and socks, helmets, and hats. A range of accessories are visible. Both machines have double hollow forks.

from his saddle. Both bicycles have alarm bells. Of course, the photograph itself is an accessory. The costliest accessory worn by these three Ottawa cyclists, however, is the outfit. During the era of high bicycles, cycling outfits varied mainly between club uniforms and much lighter racing clothes, such as those worn by Mr Laliberté in figure 2.3. In every town with a bicycling club there were tailors who catered to this trade.

Club uniforms in Canada were rather heavy, formal costumes, generally made of blue or grey woollen cloth, with braid around the sleeves and collar of the tunic and across its front, down the side of the breeches, and round the hat. In the mid-1880s, for example, members of the Toronto Bicycle Club wore a navy woollen cloth uniform with crimson frogs on the sleeves and front and crimson braid elsewhere. It may have been comfortable for a club smoker in winter, but members must have sweltered in parades on hot summer days. The uniform was intended to be an eye-catcher – riders of high bicycles did not wish to pedal by unnoticed!

A very different range of accessories is displayed in figure 4.4 by Mr W.G. Ross of the Montreal Bicycle Club. Mr Ross was one of the stellar riders of Montreal, and was a winner of numerous races. Thirty-one medals are pinned to an apron worn over his club tunic. These medals were hand-crafted by jewellers and are castings either made by the lost-wax method or fretted out from sheet metal and engraved. Most of the medals have exquisite detail. On the second medal from the right in the top row a highwheeler is cast, and on the third from the left in the next row there is the winged wheel logo of the Canadian Association of Wheelmen. Even more impressive is the silver trophy beside his right elbow. It is clear that a great deal of money was spent by race organizers and race sponsors on commissioning these medals and trophies.

Not all bicycle accessories were visible. In 1874 a bicyclist and inventor named Charles Bennett decided to tackle the delicate problem that men faced as they were jounced on their penny farthings, particularly when their pants caught in the wheel and jammed in the head of the forks, precipitating a painful header. He came up with the Bike Web, a knit and elastic garment, which, because it was worn by bicycle jockeys, was soon given the colloquial name 'a jockey strap,' subsequently shortened to 'jock strap.'[6]

Figure 4.4. Mr W.G. Ross of the Montreal Bicycle Club, photographed in 1885 with his collection of racing medals. The magnificent trophy on the table beside him has maple leaves in relief on the cup and a highwheeler set on the base.

Figure 4.5. This stand was used by William Notman in his photographic studio to support highwheeler riders. The stand itself could be concealed by vegetation, or by the rider himself, but one can sometimes detect shadows where it has been painted out on the negative. This photo is of Mr R. Jacobs, and since the date is 1892, he was one of the last members of the Montreal Bicycle Club to ride an ordinary.

Finally, some highwheeler accessories took the form of services rather than goods. Velocipede rinks and riding schools were opened in several Canadian towns (see figure 1.8). Hotels and inns advertised their accommodations for cyclists. Magazines kept enthusiasts abreast with race results and developments in bicycling. Insurers offered coverage against theft. Patent medicines warding off every conceivable malady and all forms of stiffness and soreness were offered for sale. Photographers set up studio scenes suitable for bicyclists, such as the sylvan setting created by Pittaway and Jarvis of Ottawa in figure 4.3. Notman of Montreal developed a stand shown in figure 4.5 to keep cyclists upright while being photographed (the stand was then painted out on the negative).

BICYCLE ACCESSORIES: THE SAFETY BICYCLE PHASE

In the 1890s a major expansion of bicycling took place, triggered by the development of the safety bicycle and the growing social acceptance of women cyclists. This was accompanied by a blossoming of the accessory industry seeking to reap a profit by providing a wide range of new products and services. One way of grasping the range of goods and services offered is to use, with slight modification, the classification developed by Nicholas Oddy.[7] He identifies four categories, to which a fifth is added here:

1. basic accessories that are not essential to the machine, but have high utility and may sometimes be mandatory (e.g., bells, pumps, tool kits, lights);
2. add-ons of lower utility, such as watches, cyclometers, bicycle cameras, and direction indicators;
3. components, usually replacing standard items, such as brakes, saddles, and gears;
4. items not fitted to the machine, such as clothing, bugles, and maps.
5. services to cyclists.

The first three categories – basic accessories, add-ons, and components – are illustrated using material drawn from the 1898 catalogue of the

Welland Vale Manufacturing Company of St Catharines, which provides a good indication of the growing importance of accessories. Welland Vale, manufacturer of the Perfect bicycle, chose by the late 1890s to publish two catalogues, one listing its bicycles and the other, a separate twenty-four-page catalogue, listing 'Bicycle Sundries and Accessories.' The range of accessories is sufficiently large that on the last page of the catalogue there is an index listing 64 main items alphabetically, with a wide selection of choices available under most of the main items. In total, over 180 accessories are listed. Figure 4.6 is an illustration of the bicycle oil, oilers, and pedals shown on page 16 of this catalogue. The purchaser is offered some choice: for instance, oil is sold in two different sizes of bottle, and two different shapes of oiler are available. A sense of the variety of accessories offered for sale in this Perfect catalogue can be gained by grouping them around four main subjects:

Tires: cement for rims, pumps, tire plugs, puncture repair kits, rubber patches, soapstone (to stop tires from becoming sticky), tire tape, tubes, valves, steel wool (to roughen inner tubes and make patches stick)
Lighting: calcium carbide, lamp brackets, lamps
Fixtures: bells, bags, carriers, cyclometers, coaster brakes, grips, chain guards, skirt and trouser guards, locks, toe clips, stands
Repair and maintenance: paint brushes, enamel, ball-bearings, chains, cotter pins, cranks, fork tips, handlebars, hubs, nipples, nipple grips, spokes, rims, oil, oilers, seat posts, saddles, spelter (for brazing), and tools.

Among Oddy's fourth category – items not fitted to the machine – clothing was a cyclist's most costly accessory.[8] A few items of apparel appear in the catalogues of bicycle distributors, but clothing was more commonly supplied by tailors and seamstresses. Tailor-made club uniforms seem to have remained the norm for men in Canada into the hard-tire safety phase (see, e.g., figure 2.5 taken in 1889), but during the early 1890s male cyclists began to sport more casual outfits and in summer much lighter and cooler clothes. This shift in clothing was not unique to bicyclists; but rather, it was part of a much larger transformation in social attitudes. Illustrations of men's cycling clothing in the 1890s

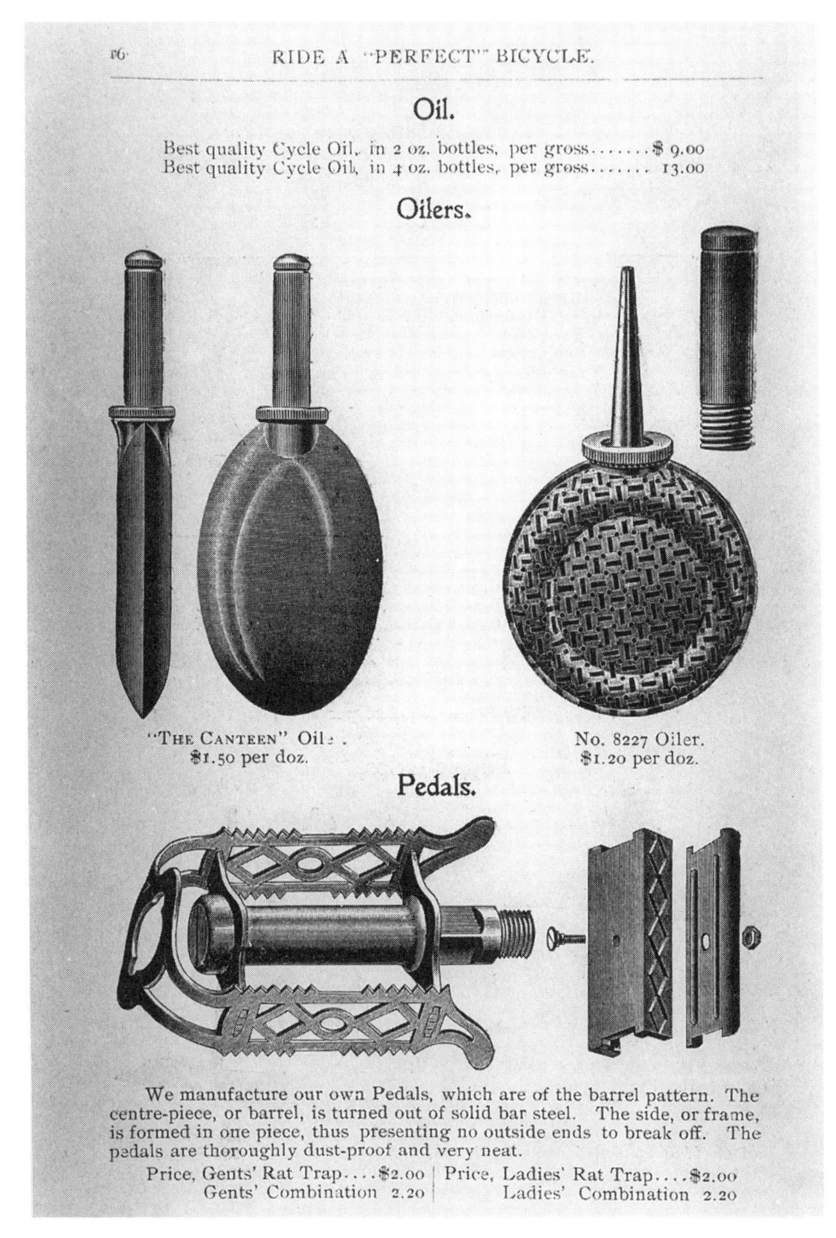
16 RIDE A "PERFECT" BICYCLE.

Oil.

Best quality Cycle Oil, in 2 oz. bottles, per gross........$ 9.00
Best quality Cycle Oil, in 4 oz. bottles, per gross........ 13.00

Oilers.

"THE CANTEEN" Oiler.
$1.50 per doz.

No. 8227 Oiler.
$1.20 per doz.

Pedals.

We manufacture our own Pedals, which are of the barrel pattern. The centre-piece, or barrel, is turned out of solid bar steel. The side, or frame, is formed in one piece, thus presenting no outside ends to break off. The pedals are thoroughly dust-proof and very neat.

Price, Gents' Rat Trap....$2.00	Price, Ladies' Rat Trap....$2.00
Gents' Combination 2.20	Ladies' Combination 2.20

Figure 4.6. A few of the accessories and sundries advertised in the 1898 catalogue of the Welland Vale Manufacturing Company of St Catharines, Ontario. At $13 for a gross (i.e., 144) of four oz. bottles, bicycle oil cost about 12 cents per bottle. The pedals were made by Welland Vale in one of a large number of styles available in the 1890s. Standardized crank threads made this diversity possible.

include the fairly casual garb worn by men on the 1896 Tour of Tours shown in figure 2.6 and Paul Pequegnat's sartorial outfit shown in figure 3.4 (the bicycle suggests that this image was taken around 1890). An enthusiastic male cyclist might have as many as three riding outfits: formal club uniforms were expected for parades, at least until the early 1890s; a more casual outfit with long socks, breeches, and jacket would be worn on regular rides; and light cotton outfits became popular for hot summer days (figure 4.7). Racers would also have a racing outfit, such as that worn Joseph Drury in 1895 (figure 4.8).

Cycling became a popular pastime for women in the 1890s, with clothing styles becoming less formal as the decade passed by. There is, however, an exception to every rule. Figure 4.9 shows Mrs McCormick of the Montreal Cycling Club in 1895 wearing a woman's club uniform. This is a rare image, showing a woman dressed in the complement to the formal man's uniform of the highwheeler phase: she wears a club hat with badge, black gloves, and a long-waisted dress uniform with leg-of-mutton sleeves and a skirt that covers her ankles. Such a long skirt requires two other accessories that are just visible on the bicycle, a skirt guard on the rear fender and a chain guard (skirt-lifters were also sometimes worn). Also clearly visible are a pea bell, a cyclometer for recording distances (attached to the front wheel hub), and some ribbons indicating that Mrs McCormick had participated in various club events. Rather puzzling are the coaster pegs attached to the front forks; although she was evidently a keen and accomplished rider, one wonders whether such a formal and proper lady would be seen with her feet up on the pegs.

More commonly, women wore elegant outfits when cycling, some quite elaborate and others more informal. Miss Kennedy was photographed in September 1897 (figure 4.10) wearing a popular style of straw boater with ribbons and flowers around it and a two-piece costume with a skirt that ends well above the ankle. In warmer weather the jacket might be dispensed with; a long-sleeved blouse and skirt would then suffice, but the hat seemed to remain a fixture. This informality was related to a much larger social transformation and should not be attributed only to the bicycle. In Anita Rush's estimation, 'the bicycle was only a small part of a much larger concentration of forces that reshaped feminine garb.'[9]

Figure 4.7. Two cyclists examine a bicycle map. Note the informal linen bicycle outfits worn by Mr J. Reilly and friend (the date is 31 August 1896). They are still wearing caps and ties, but the ensemble is much more casual than the formal outfits worn only five years earlier.

Figure 4.8. Joseph Drury of Montreal wears his racing outfit: wheelman vest, long shorts, and shoes without socks. His 1895 model bicycle has a large frame, drop handlebars, pedal clips on the pedals, and a fixed gear with no brakes. The rims are made of wood for lightness.

Figure 4.9. Mrs A.S. McCormick wears the woman's 1895 uniform of the Montreal Bicycle Club. The very high handlebars would require Mrs McCormick to ride with an upright posture.

Figure 4.10. This photograph, taken in Notman's Montreal studio on 14 September 1897, provides an interesting contrast with figure 4.9. Miss Kennedy's bicycle is a Columbia ladies safety.

These forces resulted in greater informality not only in cycling, but in many other aspects of everyday affairs: the decade was to become known as the 'gay nineties' precisely because the formalities of the mid-Victorian era were being relaxed.

What did Canadian children wear when riding their bicycles and tricycles? Very few rode highwheelers, but one image that survives is of Master Lane of Montreal, offspring of one of the city's leading families, mounted on a child's ordinary (figure 4.11).[10] Later images of children riding tricycles and bicycles indicate that they were permitted to dress with a degree of informality, at least by the 1890s. In figure 4.12 Miss L.O. Neils, who appears to be about ten years old, is shown on a tiller tricycle. She is wearing a long, smock dress with lace collar and a jaunty boy's cap.

Some accessories to bicycling lay completely outside the activity of riding but nevertheless capitalized on the popularity of the sport. Bicycle books and magazines appeared in large numbers, contributors waxing eloquent about rides to exotic places, about the health-giving properties of riding, and about how to maintain bicycles. Bicycle maps, such as that being examined in figure 4.7, became essential for tours. Eastman of Rochester made the Poco camera to mount on bicycle handlebars, and several other camera makers followed suit. Bicycle songs appeared on sheet music and bicycle photographs in family albums. A wide range of bicycle ephemera was sold in stores – bicycling images appeared on German beer steins, meerschaum pipes, crockery, picnic sets, beer bottles, ink stands, cigar boxes, and leather ware. In short, the bicycle became deeply implicated in consumption and advertising.

Finally, there were a number of accessory services provided to cyclists. Around Toronto and Montreal, tearooms became popular destinations, as did hotels catering to the cycling tourist (especially in the Niagara region). Photographers did a good business with cyclists; smaller 'cartes de visite' and larger 'cabinet' photos were much in demand by cyclists wishing to inform those who had not yet seen their latest acquisition. In these service activities, Canada was following the lead of other countries caught up in the excitement of cycling. There was one service, however, that became more important in Canada – the bicycle livery. Because of the long

Figure 4.11. (opposite) Master Lane, whose father was a leading member of the Montreal Bicycle Club, dressed in a club uniform and mounted on a child's ordinary bicycle in October 1885. The machine may be a Youth's Columbia with a 28-inch front wheel, introduced in 1880 and priced at $55.

Figure 4.12. (above) Miss L.O. Neils of Montreal on her tiller tricycle in July 1894. The tiller handle allowed a child to steer, or to be pulled along by a parent. She is dressed in a long unshaped smock dress and a boy's cap. These tricycles were popular with children in the 1890s.

northern winter, many cyclists would put their machines in the care of bicycle liveries after the Canadian Thanksgiving (mid-October), where they would be cleaned, serviced, oiled, and stored until spring. This service, too, was promoted by advertising.

ADVERTISING BICYCLE ACCESSORIES

It is simply impossible in a book with black and white illustrations to do justice to the wonderful images used in many bicycle advertisements during the 1890s; the interested reader should examine Pryor Dodge's book, *The Bicycle*, which has dozens of superb colour reproductions of bicycle and tricycle advertisements that illustrate the artistic talent put to work on promoting bicycles and their accessories, especially in France.[11] Stephenson and McNaught's collection of bicycle advertisements is another revealing source with a stronger American component.[12]

Each country seems to have developed its own artistic styles to promote bicycle sales. The French, for example, frequently used images of Nike or other minimally dressed goddesses to promote cycles and accessories.[13] The British were more functional in their approach: riders are often shown pedalling happily through the countryside – the rural idyll – complete, of course, with various accoutrements. The dominant iconography of American advertisements seems to rest on competition: the advertised marque or accessory makes its owner go faster or in some other way stand out from the crowd.

Canadian bicycle and accessory advertisements were very focused on the product itself, often with little embellishment or context. They depended more on text than on image. They were, it has to be admitted, pragmatic and restrained – and sometimes even dull.[14] In the early 1890s some advertisements had no images, simply text, although most, such as the Non-Splitable Bowmanville Rim of figure 4.13, would include a pen-and-ink drawing of the product and invoke science as proof of its qualities. These restrained advertisements formed part of a national psyche moulded by many factors, such as the Protestant values of many immigrants, including entrepreneurs such as Flavelle, Massey, and Eaton; the industrious lifestyles of a predominantly rural population; and the

Figure 4.13. Non-splitable rims, as the three big nails through the rim are intended to demonstrate. The manufacturer, Bowmanville Cycle Wood Rim Company, was located about 50 miles east of Toronto on Lake Ontario where it had good access to the hardwoods – elm, maple, and black walnut – needed to manufacture the rims. Advertisements with accurate illustrations and text were the most common type of Canadian advertisement during the 1890s.

importance of frugality and self-reliance to an immigrant society. In addition, the small size of the Canadian urban market made elaborate colour illustrations a costly proposition. A few lighter illustrations do appear, however, including the whimsical image promoting Comet bicycles shown in figure 4.14. Perhaps the most imaginative bicycling advertisement seen in Canada is found in figure 4.15: the side of an elephant in a Saint John circus parade is used by a bicycle academy to announce its services – riding lessons, cycle repairs, and bicycle sales.[15] The frequency of Canadian bicycle advertisements increased during the bicycle craze of

Figure 4.14. This whimsical and untypical 1892 cartoon advertising Comet bicycles, which were made in Toronto, was accompanied by an image of a Comet bicycle and several lines of text.

Figure 4.15. A circus parade on King Street in Saint John in 1896 or 1897 provides an interesting advertising opportunity. A young African elephant (an animal rarely found in a circus) is sufficiently eye-catching that a local firm rents the space to urge spectators to go to Burnham and Marsh's Bicycle Academy for riding lessons, repairs, rentals, and to purchase Columbia and Waverley bicycles.

the late 1890s, as it did in the United States, where bicycle advertising accounted for 10 per cent of all paid space in periodicals in 1898.[16]

Canadian sales catalogues were generally more imaginative and colourful than magazine advertisements. Shown in figure 4.16, the cover from the Goold Bicycle Company's Red Bird catalogue of 1898 promotes one vision of modernity. The old folks on the farm, which is presented as a rural idyll, hold their hands up in amazement as the young folks – perhaps their son and his lady friend – arrive on their Brantford bicycles, dressed in the latest bicycling outfits. Note that farm life is not presented critically: the old folks are contentedly sitting on the porch surrounded by

Figure 4.16. This image from the cover of the Brantford Red Bird bicycle catalogue for 1898 (the red bird is symbolized by the rooster in the foreground) presents an interesting allegory of modernity. The old couple on the farm marvel at the visible success of the young couple who have ridden up from the city. The immaculately dressed urbanites evidently personify modern life.

flowers, while the farmhand at the gate is smiling. The red bird in the foreground is happily crowing. The subtext is that modern urban bicyclists have the best of both worlds. They can take advantage of a new geography by simultaneously living in the modern city and, as bicyclists, visit the bucolic countryside whenever they wish.

AN ECONOMY OF SIGNS

A distinction has long been made between necessities and luxury goods. Demand for necessities is fairly inelastic and is not easily changed by promotion and advertising. In contrast, sales of luxuries like bicycles and their accessories can be increased by promotion, especially if they can be publicized as *signifiers*.[17] This brings the argument back to the central theme, because bicycles became one of the more visible *signifiers* of modernity during the 1880s and 1890s.

From this perspective, bicycling was largely a cultural activity. It was, of course, partly a sport and partly a recreation, but only a minority of Canadian adults were sufficiently committed to bicycling as a sport to continue doing it after 1900, when it went out of fashion. The bicycles, particularly the accessories, signified that a person was in the mainstream of popular culture at that time, just as other activities and artefacts serve as signifiers today.[18] Bicycling in the 1890s was accepted as an expression of popular culture and therefore attracted many young socialites. Thus, the deepest significance of bicycles and their accessories was in the domain of culture.

One way of demonstrating this point is to refer to David Perry's book, published in 1995, entitled *Bike Cult*.[19] In its 570 pages, this book comes close to realizing its claim to be the 'ultimate guide to human-powered vehicles.' In part four of the book, entitled 'Bike Culture,' a vision is presented of an environmentally friendly culture in which the bicycle plays a central role. In dress, in magazines, in advertising, and in love, he sees the bicycle creating a significant counter-culture. Perry concludes with the visionary words: 'the bicycle may be part of the answer to everything, as a vehicle that challenges people to find new paradigms in art as in life.'[20] By extension, the bicycle of 100 years ago also created a culture – not a counter-culture such as that envisioned by environmentally conscious bicyclists – but a mainstream culture interested in things modern. There was a necessary symbiosis during the bicycle boom between producers of new bicycles along with a growing list of sundries and services and an engaged group of affluent consumers eager to grasp the latest sign that they were in the vanguard of modern consumption.

CHAPTER FIVE

Bad Roads, Good Roads

Innovations make demands on infrastructure. Trains need tracks and signalling systems. Ships need docks and wharves. Electric lighting requires electricity-generating capacity and transmission lines. The recent expansion of computer networking has necessitated sophisticated telecommunication channels to carry huge increases in the flow of digital information. Conversely, inadequate infrastructure may slow the progress of a carrier wave.[1] Thus, the interest of Victorian bicyclists in improving and modernizing the infrastructure they used forms part of a larger pattern. Their zeal extended to the network of highways and byways along which they rode and the bridges and embankments they crossed. In seeking to improve roads, cyclists found willing allies among certain interest groups, while simultaneously triggering conflicts with others who contested the bicyclists' invasion of specific spaces: horse and buggy owners had frequent altercations with cyclists, and shopkeepers and pedestrians loudly protested the use by cyclists of the smoothest place to ride – the sidewalks.

In the late nineteenth century, road improvement associations were launched both in the United States and in Canada. Within these movements, the voice of the bicyclist was only one among many calling for better routeways; farmers, engineers, municipal politicians, haulage contractors, developers, and clergymen had been pressing for better roads since long before the bicycle made its debut. For a while in the 1890s, however, cyclists formed an especially vocal interest group within the various road improvement lobbies, and they helped to form Good Roads associations in both countries. Thus, the impression given in some histories of bicycling that wheelmen stood in the vanguard of the campaign for road improvement should be corrected; their main influence was felt in the 1890s, long after farming interests launched their lobby. The evidence also suggests that Canadian cyclists were less important to Canada's Good Roads movement than their American counterparts were to the U.S. movement.

Among cyclists, the initiative to form road improvement associations appears to have been taken first by the British: in January 1887 the National Cycle Union and the Cycle Tourist Club joined forces to form the Roads Improvement Association.[2] In the United States, the League of American Wheelmen's Good Roads movement mounted a vociferous and well-orchestrated campaign, beginning in October 1889, when Colonel Albert Pope, the campaigning genius behind the movement, made a speech fulminating about the poor state of many roads and demanding the inauguration of a 'national roads improvement movement.'[3] Soon after, the League's Roads Improvement Programme was formulated, leading in January 1892 to the launching of a new monthly magazine entitled *Good Roads*. On the magazine's cover the fundamental importance of roads was proclaimed with the following lofty slogan: 'The road is that physical sign or symbol by which you will understand any age or people. If they have no roads they are savages for the road is the creation of man and a type of civilized society.' The magazine pursued its goals with almost religious enthusiasm. The lead article of the first two issues, entitled 'The Gospel of Good Roads,' set the tone. Roads in Europe were frequently upheld as an example, but in the first three volumes there is no reference to Canadian roads. This omission suggests that Canadian roads at this date were not much better than American roads and therefore could not be held up as an example.

The Canadian Wheelmen's Association followed the example of the League of American Wheelmen's (LAW) Good Roads campaign in September 1891, when a committee proposed that 'the Dominion be divided into small districts, and that, so far as possible, there be placed in charge of each of these districts a Member of the Board of Officers, whose duty it will be to collect and supervise the preparation of reports on *every* road in his territory.'[4] Bicyclists in Canada then joined the chorus of voices campaigning to improve roads. Rush takes the position, however, that the Good Roads campaign of the Canadian Wheelmen's Association was less well organized than that mounted south of the border: 'In general, representations made to various levels of government throughout Canada tended to be poorly coordinated and sporadic ... There was no pan-Canadian movement, nor were there strong provincial movements that united cyclists into a cohesive pressure group.'[5] For example, Rush could find no evidence of petitions' being presented by Ottawa's bicycle clubs when Ottawa's City Council deliberated over paving Sparks Street, the city's leading shopping thoroughfare; the main advocates for paving were the owners of properties along Sparks Street. This argument should be qualified in two ways. First, wheelmen were not particularly concerned about the paving of downtown shopping routes like Sparks Street; their priority was to pave roads leading from the city into the surrounding countryside. Second, Canadian wheelmen at that time included a number of the elite who were capable of making their opinions known to municipal and provincial politicians through informal and private channels.

The key to the position argued here can be found in the following perceptive statement: 'Proof that a city was *modern* and *progressive* was found in its streets.'[6] Many groups had interests in building a modern city with paved streets. Even in the United States, Colonel Pope was leading a coalition of interests – indeed, it was only because of his masterful management of publicity that American Wheelmen played such a decisive role. In the absence of a charismatic leader in Canada, it was more obvious that the Good Roads movement united a broad coalition of interests. Within this coalition, Canada's Wheelmen were active campaigners. Thus, in the issue of *Cycling* published on 8 April 1897 there are several reports of cyclists lobbying for road improvement, including many of Toronto's cyclists who had attended a mass meeting in the cause of Good Roads at St George's Hall and members of the Vancouver club who felt that their own efforts in campaigning for Good Roads were noteworthy.

How bad were Canadian roads? That depended upon the season and

Figure 5.1. Stanley Park, Vancouver, in 1898. The park was a favourite haunt of local cyclists, but some of the descents were quite steep. Note the sign: 'Cyclists – Dangerous Road.' This couple play it safe by walking.

the jurisdiction responsible for road maintenance. Some roads were potholed nightmares, others were tolerably smooth. A few had rideable cinder paths along the edge. Some that were passable in dry weather became quagmires during the spring thaw and when it rained. For instance, in their last outing of 1890, which went to Highland Creek along the Kingston Road, members of the Toronto Bicycle Club encountered a stretch of road that was simply impassable, and the riders had to take to the fields to get past the morass.[7] When Karl Creelman rode across western Canada in 1899, he often followed the rail bed because there were so few rideable roads.[8] There were few cuttings and embankments along nineteenth-century roads; they simply followed the rise and fall of the

Figure 5.2. A cyclist crossing the new Lorne Bridge in Brantford, Ontario, circa 1881.

land, undulating far more than the graded roads that today follow the same routes. Steep downhills posed a hazard for bicyclists, although in some cases warning signs were posted (figure 5.1). Most country bridges were built of planks, which presented hidden dangers to the narrow wheels of bicycles. Yet short stretches of macadamized roads had begun to be built,[9] and even dirt roads, when well constructed, could provide an excellent all-weather riding surface, so the technology to build better roads was available. Modern bridges, such as that illustrated in figure 5.2, had also been constructed in a number of towns.[10]

Road improvement progressed faster in towns than in the countryside. Nevertheless, both settings posed dangers for the bicycle rider.

Figure 5.3. These tandem riders have reached a crossroads near Grimsby, Ontario, on a ride in May 1896. The scene is of a modern countryside: telegraph poles, a new rail fence, and the photographer's bicycle in the foreground point to some of changes under way.

In figures 5.3 and 5.4 photographs of the same bicycle tour to the Niagara region in May 1896 are shown. The first photo was taken near Grimsby, Ontario, on a country road whose surface was dirt laid over a gravel foundation. Such surfaces were quite good for bicycling in dry weather, although sandy patches could bring bicycles to an abrupt halt. A horse-drawn grader would occasionally smooth the surface of these roads. The second photo is taken outside a hotel in St Catharines where the rectangular stone-blocks (setts) must have given riders a severe shaking. An even greater hazard was presented by streetcar lines: if a bicycle wheel caught in the tracks, a rider invariably fell off. Note, however, that these photographs were of a planned tour on some of the best cycling roads in the province. Elsewhere, roads were generally worse.

Not all Canadian roads were bad. Indeed, the earliest roads built on

Figure 5.4. Hamilton Bicycle Club members on the Tour of Tours of May 1896 line up outside the Welland House Hotel in St Catharines (owned by the HBC).

Vancouver Island so impressed Colonel Albert Pope that he held them up as an example in his article of March 1892 in the *Forum* magazine: 'Those who have visited the island of Vancouver and especially the neighbourhood of Victoria, will have appreciated the fact that the high roads there are immensely superior to most of our own country. Their excellent condition is due mainly to the centralization of provincial government in British Columbia, and to the energy and talent of the late Governor Douglas, who personally directed all public improvements.'[11] In practice, not all of British Columbia's roads were as good as the Colonel imagined. In figure 5.5, a woman bicyclist on a rough, narrow road near Vancouver is shown passing a horse-drawn wagon with difficulty. A correspondent to *Cycling* in 1890 stated that Canada's roads were marginally better than those of the United States, with the added advantage of Canada's cooler

Figure 5.5. Near Vancouver, B.C., in July 1900, a young woman cyclist squeezes past a horse-drawn wagon on a rough dirt road.

summers, which led to the suggestion that more of America's 250,000 bicyclists should be encouraged to visit as tourists, especially if Canadian Customs allowed accompanied bicycles to pass free of duty.[12]

As bicycle technology improved and the machines could be ridden faster, road quality became a serious bottleneck in the quest for speed and in making bicycling safe for a larger public. North American bicyclists lobbying for better roads began to coalesce with other interest groups into a formal organization known as the Good Roads Association. This movement was organized at the state level in the United States, and the provincial level in Canada. Since the pioneering Good Roads movement in Canada was that of Ontario, the formation and activities of the Ontario Good Roads Association will be described in some detail. First, however, a picture of what roads were really like should be painted.

The number of accidents that bicyclists suffered, particularly on high bicycles but also on safeties, is testament to the horrendous state of many roads. A cyclist might do half a dozen headers off an ordinary in a day of riding, and riders occasionally broke wrists, arms, and collar-bones. The four worst hazards appear to have been large stones, which caused instant headers if not avoided; soft sand, causing slow-motion headers as the front wheel sank down; ruts that trapped the front wheel, causing a rider to fall sideways; and horse-drawn vehicles, either because their drivers were ornery and blocked bicycles, or because the horses were skittish.

THE TOURS OF KARL KRON

The most detailed description of Canadian roads during the highwheel era of the 1880s has been provided by an American Wheelman, Lyman Hotchkiss Bagg (figure 5.6). Bagg undertook two long bicycle trips in Canada, one in the Maritime provinces and the other in Ontario, in both cases keeping a meticulous record of the roads. These diaries form the basis of a book he published in 1887 entitled *Ten Thousand Miles on a Bicycle*. This is possibly the most boring book ever published. Its preface (which is 107 pages long) lists the contents, a general index, an index of places (including lists of rivers and valleys, mountain peaks, hills, islands, lakes and ponds, creeks and brooks, waterfalls, parks and squares, railroads, etc.), an index of persons mentioned in the text, plus addenda on League of American Wheelman politics, books, journalism, and his mileage in 1886.[13] There follows by 800 pages of text in which are described in withering detail every mile that this monomaniac rode. The names of the 3,000 unwitting subscribers who prepaid to receive a copy of the book are listed; their names are then re-listed, organized geographically by town and state (or province in Canada), together with codes indicating whether a subscriber was a clergyman, lawyer, physician, dentist, druggist, league consul, bugler, captain, flagman, lieutenant, president, secretary, or treasurer. Long sections are printed in a smaller font so as to squeeze in the most excruciating details. Having written the book, he then counted every word in it, informing the reader that there were 585,400 words, whereof 223,000 were in larger type, and 362,400 in fine type.

Lyman Hotchkiss Bagg engaged in his own idiosyncratic way in modernizing his world. He did so by assiduously classifying, in the Linnaean tradition, the phenomena he encountered. Born in Springfield, Massachusetts, in 1846, he attended Yale University, where as an undergraduate

Figure 5.6. Lyman Hotchkiss Bagg (Karl Kron) circa 1885.

he launched his career as a classifier by preparing an index for the first thirty-three volumes of *Yale Literary Magazine*. This numbing task completed, he moved to New York, where he wrote a column for the New York *World* entitled 'College Chronicle,' subsequently abbreviated to 'Col. Chron.,' which soon led to the pseudonym Karl Kron. For reasons that remain unclear, Bagg decided in May 1879 that he should ride 10,000 miles on his 46-inch-wheel Columbia ordinary bicycle with the serial number 234.[14] The beginning was inauspicious: when mounting to start his ride in Washington Square, he missed the pedals and fell sideways, dislocating his elbow before he had gone a yard. Evidently, he mastered the knack of riding soon thereafter, because a few weeks later he was clocking up 300 to 400 miles per week as he criss-crossed the eastern and central United States, venturing from time to time into Canada. It took Bagg five years to achieve his goal, but he succeeded on 14 April 1884, riding an extra fifty-five miles on the next day simply to make sure.

Yet for all its detail, Bagg's compendium is a priceless document that provides the clearest picture of what roads in Canada were like in the early 1880s. He also recorded the reactions of local people to the high bicycle at a time when it was still a novelty. He made two long excursions into Canada: in August 1883 he pedalled through Nova Scotia and Prince Edward Island, while in October of the same year he spent a fortnight in Ontario.

Bagg's route through the Maritime provinces is shown in figure 5.7. He sailed from Boston Harbour to Yarmouth on the southwestern tip of Nova Scotia. Having disembarked, he strode in his white riding costume past a bemused customs officer without bothering to pay duty on his bicycle, set his luggage on a train to Halifax, and mounted his high bicycle to start the five-day trip by road to the same destination. He followed the Bay of Fundy coast and the Annapolis Valley as far as Wolfville, where he crossed the peninsula to Halifax. For the first forty-seven miles, as far as Weymouth, he passed through territory settled mainly by Acadian (French-speaking) farmers; the track was rideable, although it was continuously hilly, with some grades that were long, some steep, and some rough and stony, forcing him to walk at least once. On the second day he set off on incredibly rough roads; it took him one and a half hours to cover three miles, mostly on foot. From that point the road improved and as he approached Digby it was excellent, but after that it became hilly, and late in the day he met 'a few rods of deep sand' that forced him to dismount.[15] He covered forty-four miles in about nine hours, averaging five miles per hour, which is not much faster than a brisk walking pace.

On the morning of day three the road was good, if hilly, whereas in the afternoon it deteriorated and became so sandy that he was obliged to take

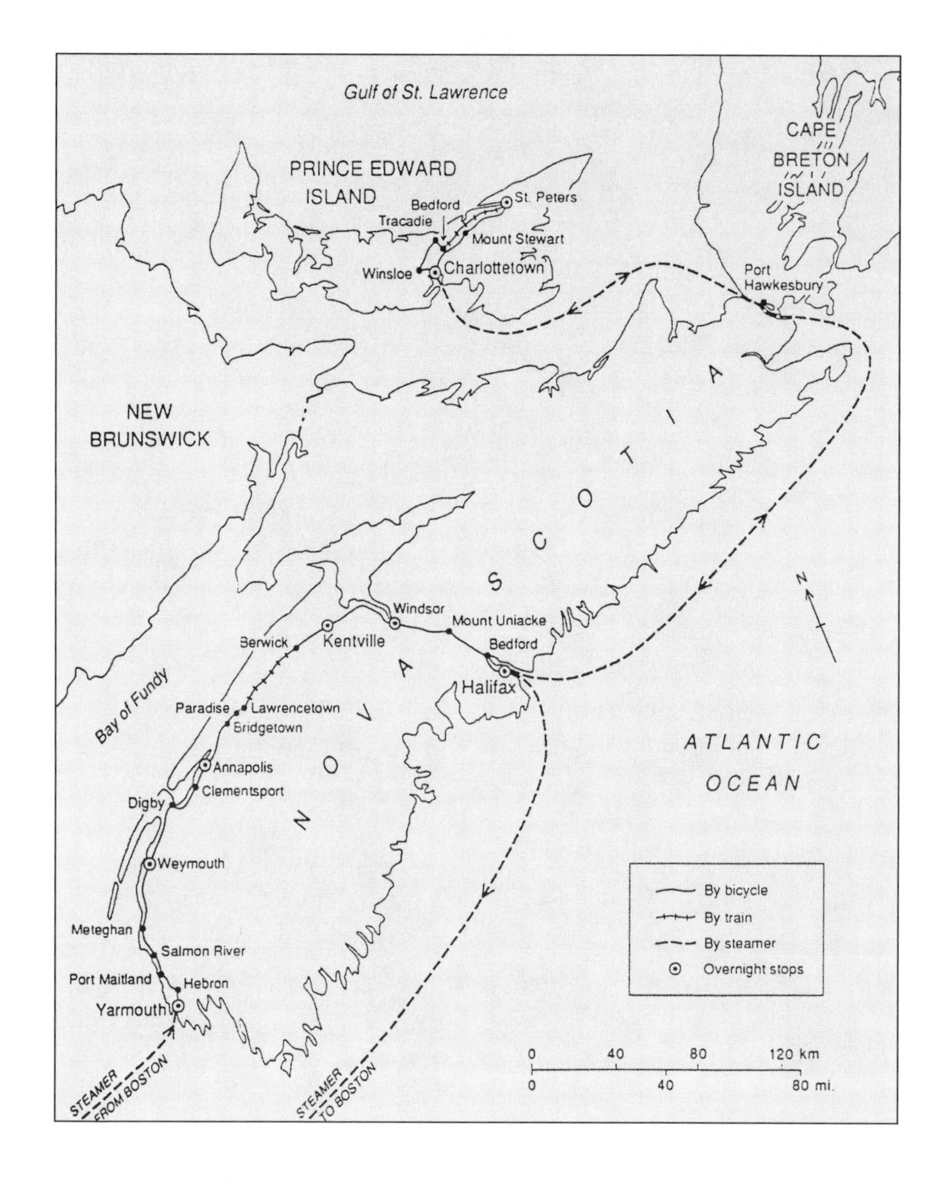

Figure 5.7. Lyman Bagg's (Kron's) route through the Maritime provinces in August 1883. He took the ferry from Boston to Yarmouth, bicycled to Halifax, then picked up a boat to Prince Edward Island, after a quick spin on Cape Breton Island. Where road conditions were too bad to ride, Kron took the train.

a train to cover the twenty-five miles to Berwick, after which he walked as much as he rode for the last twelve miles to Kentville. On the following day he rode on good roads but spent considerable time exploring the meadows reclaimed by the Acadians from the marshes of the Bay of Fundy at their historic settlement of Grand Pré. He then headed to Windsor, which he greatly approved of because 'most of its streets and outlying roads were smoothly macadamized.'[16] Day five, however, was a disaster, causing him to remark that 'if a man can live on rocks like a goat, he may settle anywhere between Windsor and Halifax.' Rain falling on the rocky road made the surface very slippery, which forced frequent dismounts, and he also had to slog through several miles of mud on the road approaching Halifax, shrouded all the time in sea-fog. He had recovered his good humour by the next day, pedalling over the city's macadamized roads in the company of local wheelmen.

Bagg might have returned to Yarmouth along the Atlantic coast of Nova Scotia, through present-day tourist meccas such as Peggy's Cove, Mahone Bay, and Lunenburg, but reports indicate that it would have been a rocky and extraordinarily difficult journey. Instead, he took a steamer from Halifax through the Strait of Canso, where the boat docked for an hour at Port Hawkesbury to discharge freight. Not one to miss an opportunity, Bagg took off on his bicycle and covered five miles on a smooth road of powdered rock. He claimed to be the first person to ride a bicycle on Cape Breton Island.[17] The steamer then continued to Charlottetown, the capital of Prince Edward Island, where his bicycle was apparently the second to be seen and the first on which the Island was toured.

From a distance the Island's roads appeared unrideable, being composed of reddish sandy clay, much rutted by heavy wagons. Nevertheless, Bagg found that, even where the ruts were bad, the grassy edges were usually firm enough when dry to allow good riding, although heavy rain could turn them to ochre-coloured mud within minutes. A southwest gale was blowing, so Bagg set off downwind from Charlottetown, along the banks of the Hillsborough River towards Mount Stewart, quite probably along the road in figure 5.8. Despite the tailwind, he averaged only four miles per hour, an indication of the state of the road and the need to walk on a number of occasions. He made similar progress in the afternoon to

Figure 5.8. This is arguably one of the finest rural landscape photographs ever taken in Canada (circa 1897). The rolling land lies north of Charlottetown; Lyman Bagg (Kron) may well have ridden his highwheeler along this road fourteen years earlier. It is easy to imagine what this country road was like during the spring thaw. Wagons used the ford, while lighter vehicles and bicycles used the low bridge.

reach St Peters, where he discovered that the hotel in his guidebook no longer existed, and he was finally 'received' at a boarding house near the station. On the following day a stiff headwind obliged him to walk frequently, and only as he returned close to Charlottetown was he able to 'wheel' with ease. Overall, he judged the roads of the Island to be monotonous, because they were laid out in perfectly straight lines, although the undulating character of the country afforded pleasant relief. On this day, and throughout his trip: 'Whenever I dismounted to quiet the fears of nervous horses the owners thereof always apologized for the trouble they had caused me, and berated their beasts for the foolishness of taking offense at the appearance of so fine and beautiful a [bicycle].'[18] In true Bagg fashion, he concludes by detailing the minutiae of this trip. By the time his steamer returned him to Boston, he had recorded 349 miles of wheeling, he had travelled 1,270 miles by boat, and 50 miles by train, and the entire expense was something less than $50.

His appetite whetted by his first ride in maritime Canada, Lyman Bagg returned to the 'Queen's Dominion' on 9 October 1883 for a second visit, this time to Ontario, attracted by 'the smoothness of these Canadian roads that ... offer a better chance than any other for testing my ability to push a 46 inch, cone-bearing bicycle straight through the country for as much as a hundred miles in the course of a single day.'[19] Bagg set off from Windsor on his tour of Ontario at 4 o'clock in the morning, heading south and east towards the north shore of Lake Erie (see figure 5.9). From there he followed the lakeshore east to Leamington, which he reached at 11 a.m. having covered forty miles – his record distance for that time of the morning. At 1:30 p.m., after a leisurely dinner break, he rode east a few miles, stopping for a brisk swim in the lake. He spent the night at a small tavern in Dealtown, having covered seventy-two miles in the day. Indeed, Bagg felt so refreshed after supper and so inspired by the bright moonshine outside, he was tempted to pack up and pedal through the evening to Clearville, which would have made a century (one hundred miles) for that day. It is a moot point whether such an achievement would have been the result of Ontario's superior roads, or of riding twenty hours out of twenty-four in a day.

Our hero rose late the next morning and between 8:30 and 10:30 a.m. had a 'smooth spin' of fourteen miles to Troy, where the village blacksmith insisted he join him for an early dinner before riding on, accompanied by his 'boy,' to Clearville. It was quite common for smiths to make crude copies of manufactured ordinary bicycles at that time, and this lad may well have ridden just such a machine while accompanying Bagg. The two riders were briefly joined by a third bicyclist in the village of Morpeth, indicating that in 1883 bicycles were not a complete novelty in southwest Ontario.

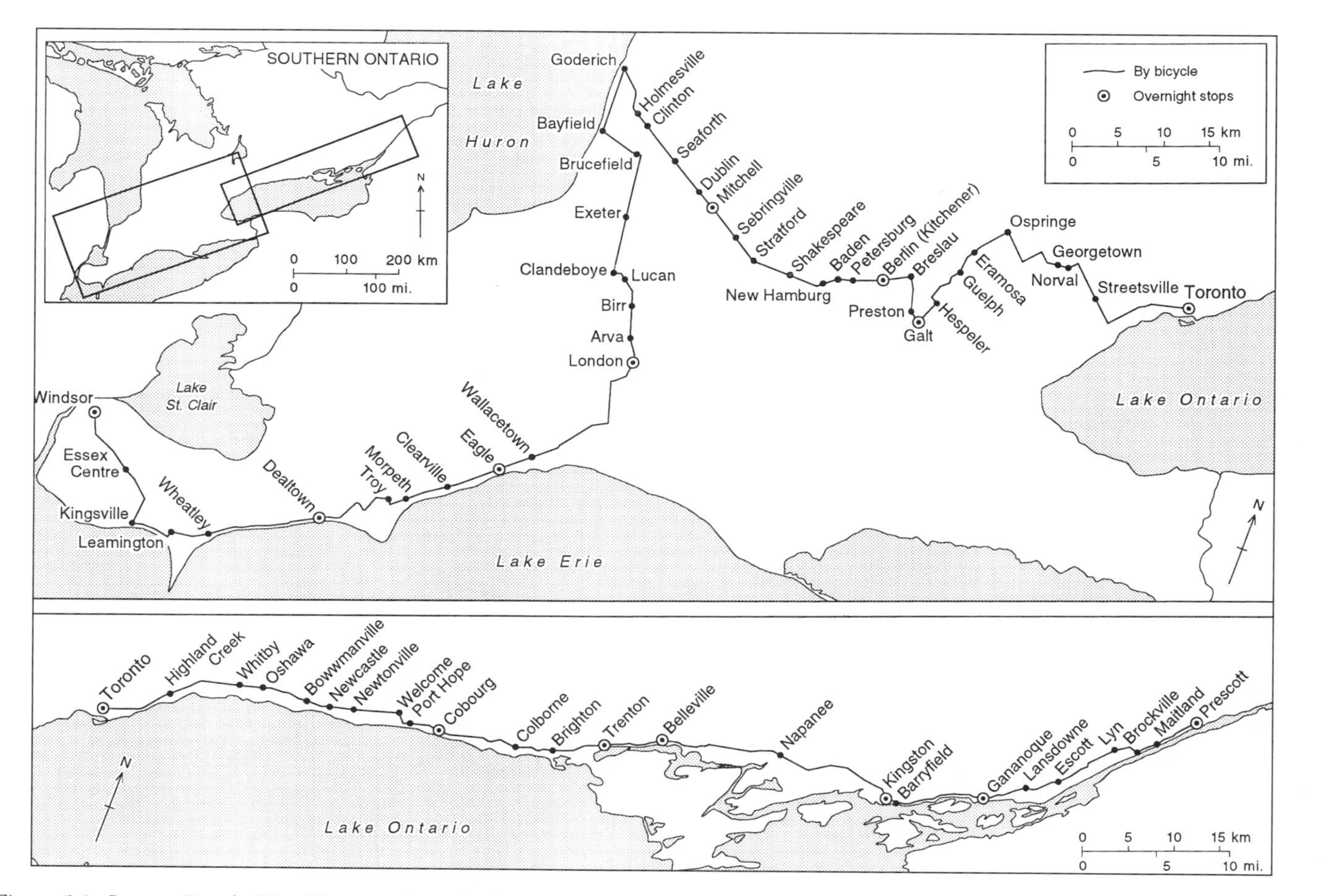

Figure 5.9. Lyman Bagg's (Kron's) route through Ontario in October 1883. Bagg completed a century (100 miles) on more than one day, though he did quite a bit of his riding at night, aided by moonlight and frosts, which showed up the pattern of the road.

At this point our chronicler launches into a lengthy aside on the wondrous state of the roads in Essex and Kent counties. He writes that, beginning at Tecumseh, eight miles east of Windsor, there is a one-hundred-mile straightaway to Clearville: 'The whole distance is practically level (i.e., there are no grades steep or long enough to be troublesome), and, when the surface is at its best, I do not think there is a rod of it which would force a good rider to dismount. In all the 6,000 miles of roadway explored by me, I know of no other stretch of 100 miles so suitable for a straightaway race; and I am sure that a fast rider who was favored by the wind might speed along this route from Tecumseh to Clearville with surprising swiftness.'[20]

From this point of his tour, the quality of Ontario's roads deteriorated. On the second day the tally of thirty-seven miles indicates how much walking was done. His account of day four is worth summarizing, since he covered a century, revealing in the process the distances a determined rider could achieve, and on what kind of diet. Rising at 5 a.m. he ate a few victuals before setting off from London at first light to ride twenty miles to Clandeboye, eating half a pound of grapes and gnawing on pieces of chocolate as he wheeled along. Breakfast at Clandeboye consisted of 'chocolate, eggs, bread, milk, applesauce and water' (312). He then rode ten miles, without dismounting, to Exeter, where he imbibed two lemonades. The good road – 'I met no other such ideally smooth and level stretch in the whole 1400 m. of my journey' (313) – continued all the way to Bayfield; he covered the twenty-two miles in two hours and twenty-two minutes (a record straightaway). After a forty-minute halt for a quart of milk, he set off on a much muddier and more difficult road for Goderich. Another quart of milk and he was off, this time heading southeast to Holmesville for a bath and for an evening meal similar to his breakfast. Assured that there was a good hotel twenty miles away in Mitchell, and good roads en route, he set off by moonlight, intent on completing a full century that day. Before he reached Clinton the moon had passed behind clouds and a strong headwind came up. Buoyed by a bottle of ginger ale, he headed on to Seaforth, where he consumed two more bottles of ginger ale before setting off at 11:15 p.m. in total darkness for Mitchell. The road was rougher than promised, and his bike pitched wildly through potholes and over stones, frequently forcing him to walk. After Dublin the road improved, and he was able to ride most of the last six miles, arriving in Mitchell at 2 a.m. as rain began to fall.

In typical Bagg fashion, he then launches into a three-page footnote summarizing the experiences of every other American wheelman who had ridden in Ontario. As usual, the details are excruciating, but there are some nuggets worth noting. Yonge Street, the principal road heading due north from Toronto into the hardrock country of the Shield, has for a century laid claim to the title of being the world's longest road. In 1883 the surface as one left Toronto was 'block pavement' and then loose macadam with sidewalks on both sides for the next five miles. Yonge Street continued with rideable hills, good coasting, and stretches of macadam as far as Barrie on Lake Simcoe, sixty miles north of Toronto.

Elsewhere in Ontario, roads did not match up to these high standards. One group of wheelmen travelling from Brantford to Hamilton reported that 'it was the "hardest" road we had yet encountered. Planks (mostly unridable, in varying stages of brokenness, filled in with unfathomable mud) formed the first 9 miles of it, and then followed a wretched "stone road," full of loose bowlders [*sic*] and ruts unrelieved by sidepaths.'[21] Other reports from bicyclists indicate that by this time Ontario had a number of macadam roads, but they tended to be fairly short sections of three to six miles, mostly in and close to towns. Elsewhere, roads varied from the rideable (gravel) to the unrideable (sand and rocky stretches). The weather was a key variable. Clay was impassable when wet, but rideable when dry if it was not too deeply rutted. Equally important was the state of the edges; wagons and carts chewed up the centre of the roadway, but bicyclists could often make good progress along a smooth verge provided that wagons had not used it.

Lyman Bagg's tour of Ontario continued to Toronto, mostly on hard gravel surfaces that were unaffected by rain, although as he pushed up the hills of the Niagara escarpment, he broke his handlebars and was forced to walk to Georgetown, where they were welded. The road into Toronto included the 'swiftest and pleasantest moonlight spin' of his experience; the road 'glistened whitely' in the frosty moonlight, making it easy to avoid ruts and holes. He does not record the reaction of the landlord of his hotel in Toronto when he arrived at 2:40 a.m. and demanded a hot bath and nourishment! He had covered eighty miles in twenty hours on this day.

During his stay in Toronto, Lyman Bagg made the acquaintance of a number of Canadian wheelmen, notably R.H. McBride, who was president of the Canadian Wheelmen's Association, and captain of the Toronto Bicycle Club. Three weeks after Bagg's departure, McBride was to set a Canadian day record by riding from Toronto to Belleville (117 miles) in fifteen and a half hours, at an average speed of seven and a half miles per hour (the first hour and the last six hours were covered in darkness). This is a good indicator of how bad the roads were.[22]

After spending a day in Toronto, which he felt was 'more wide-awake and American-like than any other Canadian city,'[23] Bagg continued his journey east along the shores of Lake Ontario, passing one of the favourite haunts of local bicyclists, the Halfway House, located about an hour out of town (see chapter 7). The roads as far as Belleville were mostly rideable; indeed the worst section was the stretch leading out of Toronto, which he deemed unrideable.

Day twelve, from Belleville to Kingston, was 'an ideal run on an always smooth road,'[24] although cool autumn weather forced him to bundle up. On the following day he experienced a short ride on smooth roads to Gananoque, while on day fourteen the ride was longer, in bitingly frosty air into the teeth of a headwind to Brockville. The last stretch, from Brockville to Prescott, was mostly on roughly frozen track! Arriving too late to catch the last ferry to Ogdensburg on the U.S. side of the St Lawrence River, Bagg had to spend one more night in 'cheap Canadian lodgings' on a couch of straw in a stuffy, kerosene-lit bedroom.

Bagg's chronicle leaves us with a good understanding of what bicyclists faced in the 1880s In winter and until roads dried out after the spring thaw, cycling was not feasible; it was largely a summer and fall activity. Small sections of macadamized roads were very popular with cyclists, who made great use of them for rides and races. Elsewhere, the quality of roads depended mainly upon bedrock and topography and the level of improvement practised by the local municipality. Clay was a disaster; it turned to glue when wet and formed concrete-like ruts when baked dry by the summer sun. Sand provided a better surface when wet, but when it was dry it could form bunker-like traps into which narrow-tired bicycles sank. Rocky areas and moraines often provided satisfactory riding terrain, because gravel pits found in such areas made it easier to improve the riding surface – provided the local municipality made the effort. Both riders and bicycles suffered frequent bangs. Bent forks, broken wheels, and twisted handlebars (and the bumps and bruises that accompanied them) were a way of life in the high-bicycle era. Riding safety bicycles was less hazardous, but compared with today's sport, there were still plenty of knocks and spills. No wonder, then, that bicyclists were vocal in their demands for improved road surfaces.

THE GOOD ROADS ASSOCIATION

Today, the Good Roads Association is a lobbying organization representing interests intent on spreading four-lane highways across the land. It is somewhat ironical, therefore, that this movement was founded partly by cyclists, a group that is now seen to occupy the slow lane, as far as such developments are concerned. In truth, the bicycle has played a paradoxical role in the drive for modernity.

The 'road problem' that bicyclists and other road users encountered in the 1890s stemmed from the way roads had been built and improved in Canada's pioneering phase. In most cases, settlers either followed the trails and portages used by indigenous peoples or cleared and maintained paths along concession lines and sideroads, but only enough to convey their farm products and lumber to the nearest waterway, which was the best means of transport and was used whenever possible. Ontario's first Highway Act (dating back to 1793), required incorporated towns and townships to appoint overseers of highways to supervise roadwork done under a labour tax known as *statute labour*. In a township requiring statutory work on roads, residents did their road-maintenance tasks under the direction of *pathmasters* of varying (mainly low) skill, generally doing as little as possible with inadequate tools and materials. This labour tax varied in proportion to the assessment on a taxpayer's property, and ranged between six and twelve days per year.[25] All persons doing such statute labour were supposed to provide their own tools. In consequence, roads maintained by statute labour were generally in bad to dreadful condition.

In 1804 the legislature had voted monies to assist in building new roads and repairing existing ones under the supervision of provincial commissioners. This action created a second and quite separate system of road maintenance. Yet a third independent power in the building and repair of roads was created in 1829 by the Road Toll Act, which permitted the formation of joint stock road toll companies to operate turnpikes. After 1853 these rudimentary roads faced increasing competition from steam locomotives operating on a rapidly expanding rail network.

By the time of the bicycle boom, two changes had occurred. Most of the road toll companies had become moribund in the face of rail competition, and their roads had been taken over by county councils. Second, quite a number of municipalities had replaced statute labour by a levy in lieu of this requirement (this process was known as commutation).[26] Townships commuted all or part of the required statute labour in return for a payment of between 30 cents and $1 per day, which then was used to pay an experienced and well-equipped gang of men working under a superintendent of roads (one report states that better results could be obtained by commuting for the meagre sum of 25 cents per day, such were the niggardly efforts of statute labourers[27]). One consequence of this patchwork system was great variability in the quality of roads from township to township, and county to county, as Bagg noticed.

In order to infuse some order into this bedlam, pressure grew in the early 1890s to upgrade roads and strive for some uniformity in their quality. Thus, in February 1891 Canadian Wheelmen were urged to use the forthcoming dominion elections to put pressure on parliamentary candidates to improve Canada's roads. The initiative to form the (Canadian) Good Roads Association came from the Canadian Institute (now the Royal Canadian Institute) in Toronto, which on 2 December 1893 passed the following resolution: 'The Canadian Institute having followed the various steps which have been taken in the past few years to awaken an interest in Road Reform, and recognizing the inestimable benefit to the community arising from improved and well constructed roads – be it resolved that the Canadian Institute issue an invitation to all persons interested in Road Reform to meet in Convention, here, at an early date for the purpose of forming a National Road Improvement Association, and that the Council be requested to make the necessary arrangements for holding said convention.'[28] True to their word, the Ontario Good Roads Association held its inaugural meeting at the Institute on 9 February 1894.

The first constitution of the association set out its organization. More interesting is the memorandum attached to the constitution listing eight 'Plans and Purposes of the Association.' These include acting as a coordinating body for all road-reform pressure groups, awakening public interest, fostering publication of road-improvement plans, establishing the broadest possible base of support, disseminating information on recent legislation and on good road-building practices to local associations, and procuring publications on the best road-building practices at reduced prices for local associations. These are normal motherhood statements for such a lobbying group, but one 'purpose' is much more specific, and revealing: 'to aid in providing for a proper road exhibit and instruction in road making at all Farmers' Institute, County, Dairymen, Creamery, Cheesemen, and other Association Meetings.' Clearly, agricultural interests were of central importance to the association.

Thus launched, the association set about its task. To make its point, a train was used by 'Good Roads' to tour the eastern part of Ontario, promoting the cause of better highways.[29] Pressure was put on townships and counties to abolish statute labour and the remaining toll roads. A year after its founding the executive of the Goods Roads Association met in Guelph in December 1895 to pass a series of important resolutions, based on their experiences during their first year of operation. The key point was the resolution that county roads should form a system connecting all local municipalities, the county level being considered the 'best adapted to advance road improvement in the Province of Ontario.'[30] This amounted to a very modern vision of road transport, since at that time most long-distance transport was by water or rail. By seeking to integrate the road system on a far bigger geographical scale than was then the case, the Association anticipated many twentieth-century developments.

Several interest groups supported the Good Roads initiative. One group clearly identified was composed of farming interests – the cheesemen, dairymen, and others who wanted to market their products. The civil engineers who would construct the roads made up another; indeed, the original motion to create the Association was moved by Alan Macdougall, a civil engineer by profession. Yet another group comprised journalists – the first president of the Association was Andrew Pattullo (editor of the *Woodstock Sentinel-Review*), and the first secretary was K.W. McKay (editor of the *Municipal World*). To these groups should be added municipal politicians conscious of the need to do something about the road problem. There were also the bicyclists: a historical note published in the fiftieth-anniversary edition of the *Municipal World* in 1944 remarks that 'urgent were the demands for good roads from the then popular bicycle associations of the day.'[31] In practice, the Association united a formidable combination of rural and urban interests; the farmers' lobby wanted good road links to cities to market their produce, while city-based bicyclists wanted good roads to gain access to the countryside.

An indication of just how interested bicyclists were in touring the countryside is provided by the many maps drawn specifically for cyclists. In figure 5.10 part of a bicyclists' map of the Toronto area is shown, printed in the 1890s. Roads marked on the map are passable by bicycle: impassable tracks and concessions are not plotted. Note that the cyclable road network is much sparser than it is today. Maple, for instance, was then accessible by bicycle only from Richmond Hill, located about four miles to the east on Yonge Street, whereas by the 1920s good roads had reached Maple from all four cardinal directions.

It is noteworthy that bicyclists did not play a major role on the executive of the Good Road Association, where economic and municipal interests dominated (in the United States, cyclists played a much more prominent, formal role). Being organized into many active clubs, bicyclists did provide a formidable lobbying group however, not least because their members were generally persons of standing within local society. Their energies were soon put to work pushing for road improvements. In figure 5.11, an 1895 membership card of the Canadian Wheelmen's Association exhorts its members to support the push for Good Roads.

These lobbying efforts brought some results. From the mid-1890s

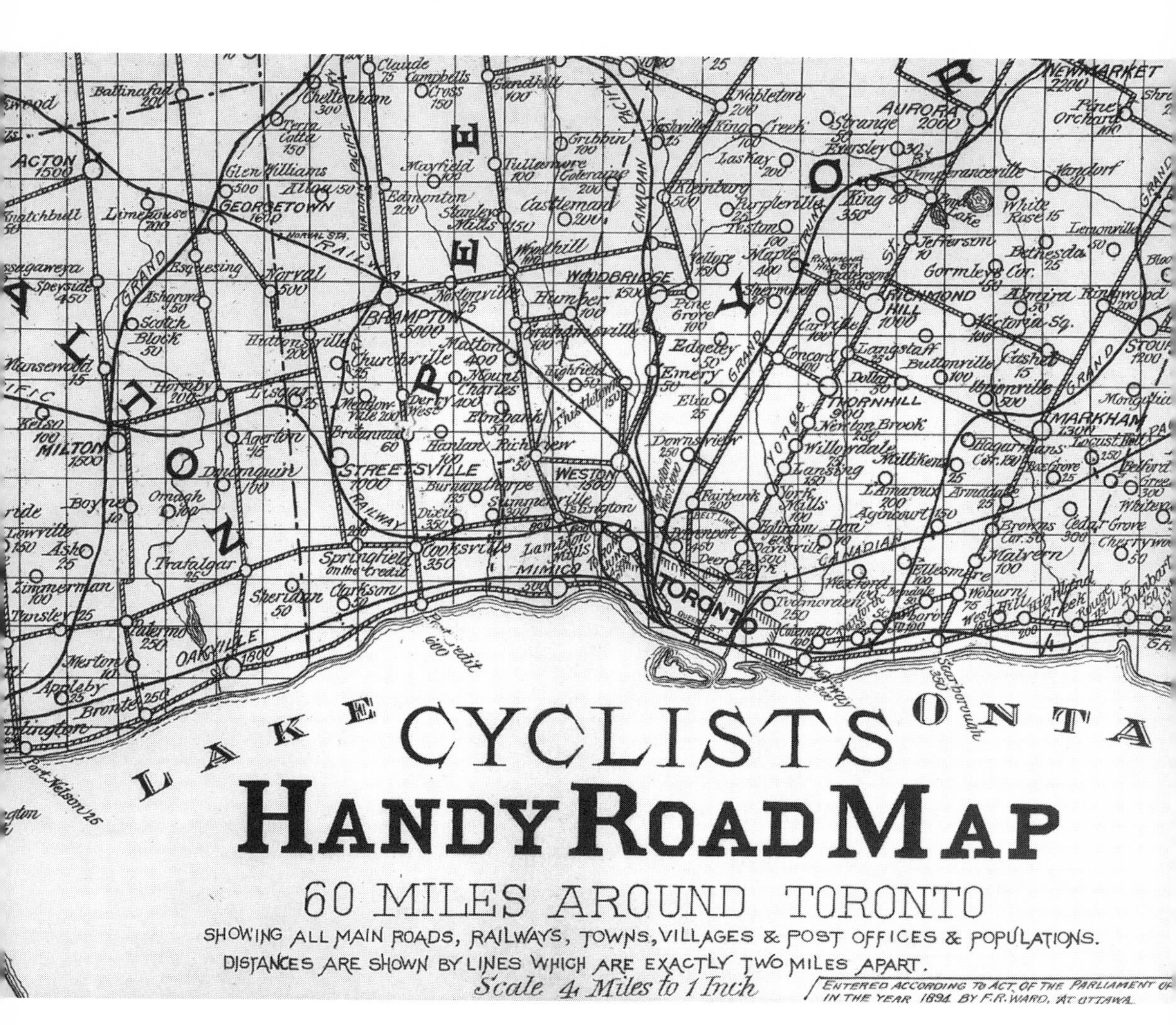

Figure 5.10. A cyclist's map of the Toronto area, 1894. The cartographers have plotted only those roads that were useable by cyclists. Thus, north of Toronto up Yonge Street there are no roads that connect westwards with the road passing north-south through Woodbridge, Kleinburg, and Nobleton. Railways are plotted because cyclists frequently took a train out to the countryside and rode home, or vice versa.

most cities made serious efforts to pave their main streets and improve the system of storm sewers so that road surfaces did not become periodically waterlogged. The following report on Winnipeg is representative of the state of most Canadian cities at this time: 'From 1895 onwards the city had engaged in a determined programme of grading and macadamizing streets. Cedar block pavements in the downtown area were replaced with new asphalt surfaces by 1899, and by 1902 most streets in the downtown were paved and construction crews were extending the pavement of streets into the emerging suburbs within the city limits.'[32]

This account of road improvement would not be complete without mention of another aspect of cycling infrastructure, namely, cycling paths. John Lehr and John Selwood have examined the case of Winnipeg, where problems arose in the 1890s as a result of cyclists' riding on sidewalks, a common occurrence when roads were in an abominable condition. They write: 'As early as 1895 concern had been expressed over the dangers of permitting bicycles on Winnipeg's sidewalks, and exception was taken to a proposal to allow bicycles on all streets except Main Street and Portage Avenue. To contain the "bicycle nuisance" a group of prominent citizens advocated passage of a by-law which demanded that: No person shall ride any bicycle upon or along any public sidewalk of the paved section of any street of the city of Winnipeg, nor upon the sidewalk of any park or the footpath of any bridge of the city, nor along the sidewalk of any street within the limits of [the downtown area].'[33] This request was accompanied by demands that speed limits of eight miles per hour be imposed on cyclists using Winnipeg streets, and six miles per hour (a little above walking pace) on sidewalks. All bicycles were to be equipped with an effective bell or alarm, and on narrow sidewalks (less than five feet wide) cyclists were to dismount before passing a pedestrian. Partly in response to these restrictions, the Winnipeg Bicycle Club initiated a network of bicycle paths. Supported by the Canadian Wheelmen's Association, the Winnipeg Bicycle Club persuaded City Council to construct a bicycle path along Portage Avenue, but it was not protected from wagons, horses, and other animals, so the carefully rolled surface was frequently damaged. Only in 1899 was 'A By-Law for the Protection of Bicycle Paths' passed by Council. The cause of Winnipeg's cyclists received a further boost in

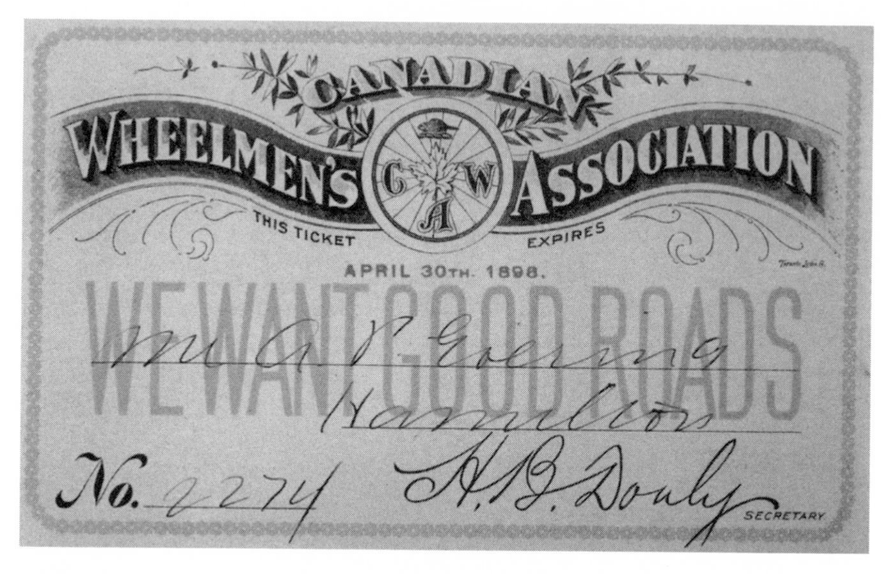

Figure 5.11. A Canadian Wheelmen's Association membership card for 1897. Signed by Hal Donly, this is the card of Mr A.P. Goering, captain and master of the Hamilton Bicycle Club and a leading figure in Canadian cycling at that time. Note the slogan carried on this card (and on cards of the succeeding years), 'We Want Good Roads.' Wheelmen provided Ontario's Good Roads Movement with a powerful lobbying force.

1901, when the Manitoba provincial legislature gave assent to a bill giving the city authority to create a Bicycle Paths Board with a mandate to establish and maintain a system of bicycle paths both within and beyond the city limits. By 1903 the Bicycle Paths Board – the first of its kind in North America – was administering some twelve miles of trails linking the downtown area with Elm Park and Silver Heights, two of the main recreational attractions close to Winnipeg. The cost of maintaining these paths was met by levying an annual licence fee of 50 cents on every bicycle ridden in Winnipeg.

CONCLUSION

In both Canada and the United States the push to modernize roads came from a coalition of interests sharing a common concern. In Canada, however, there were no charismatic leaders comparable to Colonel Albert Pope

Figure 5.12. Hunter Street at the corner of Water Street in Peterborough, around the turn of the century. The boardwalks are being replaced by concrete sidewalks. On both sides of the street men lean on their bicycles and watch workers transporting cement from a horse drawn cement mixer and pouring it into forms.

to motivate the cyclists, and as a result, cyclists did not play as prominent a role in the Good Roads movement as they did in the United States. Yet the contribution of cyclists to the Canadian campaign should not be dismissed. Despite Anita Rush's contention that bicyclists were not numerous in Canada[34], in elite commercial, financial, and political circles of the 1890s cycling had a strong following, and was often supported by non-riding parents and relatives. In Ontario, for instance, bicyclists constituted a sufficiently powerful group that in 1897 they were able to persuade the Ontario legislature to pass an act regulating the bicycle; thereafter, the bicycle was fully recognized as a vehicle, with travelling rights on Ontario's highways.[35] In the same year, the Canadian Wheelmen's Association lobbied the dominion Parliament to introduce a bill compelling railways to carry bicycles as baggage, another indicator of the bicyclists' political presence.[36]

As a result of pressure from bicyclists and other factions in the Good

Figure 5.13. Sparks Street in Ottawa on 26 June 1900. An interesting street scene just before the automobile became popular.

Roads coalition, many improvements were made to Canadian roads in the closing years of the century. In figure 5.12, for example, work in progress is shown on Hunter Street in downtown Peterborough, Ontario. The result was a significant improvement in road quality, particularly in major towns. The illustration of Sparks Street in Ottawa on 26 June 1900 gives a sense of what a well-maintained city street was like as the bicycle era came to a close (figure 5.13). The roadway is shared by bicycles, streetcars, and pedestrians; there are at least seven bicycles visible; sidewalks are made of concrete; two workmen are making minor road repairs; and the cyclist in the foreground has a smooth ride on a macadam surface.

Soon after, with the advent of motor vehicles, another powerful voice was added to the chorus for better roads. Alas, once the genie was out of the bottle, it was hard to replace it. Within a quarter of a century that flawed American genius, Robert Moses, was proving the possibilities of the expressway by carving his way through New York City and the surrounding countryside, building the Long Island Parkway.[37] In the 1930s Moses seized the opportunity presented by Roosevelt's commitment to public works to reshape American cities by building expressways, bridges, and public parks. In post-war years this model was pushed to its most destructive limits as expressways were rammed through city and countryside, thereby destroying the fabric of urban and rural neighbourhoods shaped over long periods of human occupation. Today, ironically, bicyclists are demanding the construction of bicycle paths, which remove them from the good roads for which they had once lobbied.

CHAPTER SIX

The Cycling Crowd: Modern Life on Wheels

Like most innovations, the bicycle had both negative and positive effects, triggering numerous contests over patents, markets, its use, its regulation, its safety, its medical effects, and so on. Notwithstanding these contests, the carrier wave associated with the bicycle ratcheted the project of economic modernity forwards several notches. The bicycle had a substantial impact on technological innovation, on production methods, on consumption patterns, and on road improvement.[1] Modernity is a contested project, however, and it involves more than the economy; it is a broad cultural movement that infuses many aspects of daily life. Although historians view modernity as the dominant vision of that era, there were counter-movements that attracted considerable attention. Among others, romantic poets, pre-Raphaelite artists, engravers, and painters of the sublime and the picturesque, religious traditionalists, Luddites, and racial supremacists looked elsewhere – often backwards to the past – for their inspiration; their vision, which was essentially anti-modern, created a series of alternative cultures. Indeed the bicycle – in itself a modern vehicle – paradoxically was used by many riders to escape the city and savour the bucolic pleasures of the countryside. Strands of the past and the future, the rational and the irrational, were evidently woven into the complex tapestry of modernity.

On the technical side, Macaulayite historians have stressed the beneficial consequences of the bicycle in their narratives of progress in bicycle design and construction. In practice, not all its impacts were benign; bicycles were put to all sorts of uses, some good, some bad, some sensible and some weird. They were used for leisure, work, shopping, war, racing, exercise, circus acts, fire services, public transport (as rickshaws), and daredevil stunts. Their fabrication created toxic work environments, especially in the soldering shops where tubes were brazed together; oppressive labour regimes on the proto-assembly lines; and precarious work for the salesmen travelling the countryside touting their wares. On the technical and economic sides the story is not, therefore, one of unqualified progress. The modernizing effects of the bicycle on society are also somewhat ambiguous.[2]

In this chapter the social impact of the bicycle, which, as already noted, itself is a socially constructed artefact will be examined. Thus, the bicycle is both the product of social change and an instrument causing further social change. According to the proposed carrier wave model, the major social impacts of an artefact do not occur in the early phase of intense technological innovation, when bewildering changes in product design and configuration tend to inhibit large-scale production. As product design stabilizes, so manufacturing costs and product prices decline, leading, for items such as the bicycle, telephone, and radio, to big increases in sales. The main social consequences of these innovations are then worked out during the phase of widespread adoption.[3] This is not to deny that social reactions often appear at an earlier stage – there is, for instance, already strong opposition to genetically modified foods that are still largely experimental – but the negotiation of an enduring social impact depends upon there being a critical mass of adopters. The boneshaker bicycle, for example, appeared in Canada in such small numbers and held consumers' attention for such a short period that its modernizing impulse was slight. Broadly speaking, the effects of the bicycle on Canadian society can be grouped into those of the very male-dominated phase of the highwheeler in the 1880s, and those of the more inclusive phase of the safety bicycle in the 1890s.

The social outcome of the bicycle on Canadian society is much in dispute. Take, for instance, the often-stated proposition that bicycles contributed to the liberation of women. The conventional viewpoint is outlined by Heather Watts in an essay on early cycling in Nova Scotia.

She writes: 'At a time when higher education, women's suffrage and the movement for dress reform were all topics of heated discussion, the bicycle became one more liberating influence on the restricted lifestyle of Victorian women ... This element of freedom and independence greatly appealed to women. They were no longer left at home, but could go on outings with their women friends or accompany their young man on an equal basis. Once tasted, the new freedom was hard to abandon.'[4] Rather different is the viewpoint expressed by Anita Rush: 'Both popular and academic social histories ... of the bicycle have attributed incredible achievements to it, [including] liberating women from restrictive clothing and social roles ... These extravagant claims permeate our historical literature; ... their accuracy should be questioned ... the bicycle in fact stands as a reflection of commonly held attitudes and values of the period. It was less a transforming agent than a mirror. Canadians were passive consumers of the technology. [p.1] ... One claim that should be dismissed ... is that the bicycle erased social barriers [p.2] ... The bicycle was only a small part of a much larger concentration of forces that reshaped feminine garb.'[5]

The repercussions of the bicycle are, therefore, in need of careful examination, since many of the common conceptions are contentious and the terrain evidently is treacherous. Moreover, it cannot be assumed that the social impact of the bicycle in France or Britain or the United States was replicated in Canada, since it has already been shown that other aspects of the bicycle carrier wave varied significantly among these four countries. The issue will be untangled step by step, first, by looking at the specific social categories of age, class, and gender. Broader social impacts will then be examined in the context of bicycle clubs, which provided an important focus for the social life of many Canadian cyclists, and the issue of Sunday streetcar operation in Toronto. In the concluding section we will examine the broader implications of the bicycle for social change.

YOUNG SCORCHERS? THE QUESTION OF AGE

Victorian society was typically built on principles of seniority and patriarchy (matriarchy in the case of the Queen, herself); indeed, some institutions, such as the established church and universities, came close to being gerontocracies. Cycling, by contrast, presented younger persons with the opportunity to regulate their own lives. The best evidence of the age of Canadian bicyclists in the late nineteenth century is found in several hundred photographs taken both in the studio and outdoors. These images point to a significant age difference between the riders of ordinaries and those of safeties.

Riding a high bicycle was not an old man's sport. The risks of a header and the need to be fit to mount the bicycle were two major deterrents, but the single largest barrier may well have been a social one. In the mid-Victorian period, there were very few older men who exercised vigorously. Be their sport soccer, rugby, running, or cycling, bourgeois Victorians hung up their boots at a younger age than men today because it was accepted that middle-aged men led a fairly sedentary life. A restrained game of cricket might be enjoyed, as might hunting, where the horse did most of the work. There are very few surviving images of men beyond the age of forty years riding highwheelers in Canada in the 1880s.

It should not be concluded, however, that riding ordinary bicycles was therefore the preserve of teenagers. Photographs show that many riders were well into their twenties and thirties, for two good social reasons. First, these bicycles were fairly costly; hence, unless a man was independently wealthy, he had first to establish his career before he could afford to purchase a machine, and that career had to be a remunerative one. A working man could not afford a high bicycle or the various accessories, nor did his long hours of work leave him much time for leisure; Smith notes that the cost of a highwheeler (around $100 Canadian) was equivalent to four months' wages of the average factory hand.[6] Riders, therefore, were almost exclusively of the business and professional classes. The youths riding and racing high bicycles were often the scions of leading Canadian families – such as Mr Ross in figure 4.4, whose parents presumably purchased for him a bicycle and all the appropriate accessories before he began his working career. There is a second reason that many riders were in their late twenties: middle-class Victorian men quite commonly delayed marriage and the family responsibilities that accompanied it. An indicator of the age of highwheel riders can be found in figure 6.1, in which are shown six men photographed in about 1887 on the pier at Collingwood, Ontario. None of these members of the First Collingwood

Figure 6.1. A rare outdoor photograph of high bicycles taken about 1887. Members of the First Collingwood and Stayner Bicycle Club gather on the pier at Collingwood. The riders are identified as George Mathers of Stayner; the Reverend D.H. Currie, H. Fanjay, and John Leary of Collingwood; Will Petrie and Ernest Woods of Stayner.

and Stayner Bicycle Club was a youth, and most appear to be closer to thirty. The gentleman in the bowler hat second from the left was the Rev. D.H. Currie, a clergyman who had been ordained and was launched in his Collingwood ministry at this time.

Riding safety bicycles was a sport that attracted a much broader age spectrum. A comparison of figures 6.2 and 6.3 illustrates this point; both photographs were taken at the same location outside the New Brunswick legislature in Fredericton, the former during the Bluenose Tour of 1886, the latter circa 1896. Many changes occurred during the intervening decade, but for the present, age is the issue. Most of the riders in the earlier photograph are in their twenties, although the riders of the two tricycles and a couple of the highwheel riders may have reached thirty. Very different is the age of the riders in the later picture. In addition to the portly gentleman in tailcoat and straw boater at front centre, there are a couple of older, bewhiskered gentlemen in their fifties or sixties. Equally important, children have appeared – a dozen are visible. Thus, the safety bicycle opened up cycling to all but the very elderly. This broadening in the age range of bicyclists was a major factor in expanding the demand for bicycles in the 1890s.[7]

Figure 6.2. The Elwell Bluenose Tour of 1886 in front of the New Brunswick Legislative Assembly, Fredericton. Notice the two-track tricycle (left) and three-track tricycle (right), the elaborate pose with two pyramids of cyclists, and the brace of five bicyclists on the top step. There are twenty-four high bicycles, two tricycles, and thirty-three persons (all men); that is, all but seven persons have a machine in the picture. Most of the men appear to be aged between twenty and thirty.

Figure 6.3. The Fredericton Bicycle Club in front of the Legislative Assembly building, circa 1896. In the ten years since the previous photo was taken, some significant changes have taken place. The age of cyclists has broadened dramatically to include riders from about six to sixty years old. Bicycling has become a family affair: women and children appear in the picture. The high bicycle has disappeared completely. All these machines appear to be pneumatic safeties. But most important, the social side has claimed some ascendency over the bicycles themselves. There are about sixty-seven persons and a dog in the picture, but only about twenty-eight bicycles, which are now less prominently displayed.

Figure 6.4. The generation gap.

When the term 'scorcher' was first used is unclear, but the word was very much in vogue during the 1890s to describe fast riders. A scorcher was implicitly young; indeed, the term 'old scorcher' would seem to be an oxymoron.[8] Scorching was not associated with highwheelers (except in races); first, because the machines were lethal enough without extra risks, second, because the military style of many club rides required conformity; and third, because few highwheel riders were youths – the machines cost too much. This is the key point: most of the scorchers who were the object of critical comment in newspapers and in the newsletters of Canadian bicycle clubs were in their late teens and early twenties, the very same age group frequently accused of driving automobiles recklessly a century later. It was rising incomes, declining bicycle prices, and the diminishing discipline of club rides in the 1890s that enabled large numbers of youths to purchase a bicycle and then go scorching with impunity. These young scorchers became such a problem that the Toronto Bicycle Club passed a motion in 1892 insisting on 'No scorching on Club runs.'[9]

What did their parents think? The response was no doubt the same as it is today when teenagers ask their parents for the keys to the family car. Some were indulgent and bought their children bicycles; others disapproved. Figure 6.4 is a difficult photograph to 'read,' but it would seem to be a commentary on the generation gap created by bicycling. Two men in their thirties deliberately pose with their new bicycles. The older gentleman, probably their father, stares fixedly in another direction as if disinterested in the whole affair, while the young woman looks on with curiosity.

THE CYCLING CLASS

There are grounds for arguing that cycling had a bigger influence on social modernity through its class relations than through its gender relations, in essence because it served for much of the bicycle era as a highly visible status symbol. Andrew Ritchie is one of the few authors to have examined this question.[10] Ritchie explores the difference between the 'gentlemen amateurs' of British bicycle racing, like the Hon. Ion Keith-Falconer (an aristocrat and undergraduate of Cambridge University), and the 'working professionals' like John Keen. Given the virtual absence, in the early days of cycling, of a group of professional racers in Canada, the issue of the bicycling class should be examined in a slightly different frame.

In Canada, the fundamental class relation of cycling was very simple: through both its highwheeler phase and most of its safety bicycle phase, bicycling was the preserve mainly of the Anglo elite. The bicycle became a signifier of a person's status. The nominal cause of the bicycle boom's ending in the late 1890s was the collapse of sales, but this was a temporary setback; as was shown in chapter 3, sales of bicycles recovered to fairly high levels in the early 1900s, but the 'boom' was never revived. The argument made here is that revival was not easy, because the boom was a social relation defined more by the participation of Canada's elite than by the actual number of machines sold. Its essence was the association with the project of modernity that was directed mainly by the Anglo upper crust. The succession of novelties during the boom – new bicycles, new accessories, and new cycling experiences – fuelled the interest of the elite in cycling. By 1900 a major decline in the number of bicycle-related novelties was apparent, as technological innovation switched to other fields. Sales of mass-produced bicycles would rise again, but they were the no longer positional goods that enticed the elites of Canada. When their interest turned else where, the bicycle lost its social cachet.[11]

Some evidence of the Anglo elite's role in cycling is found in figure 6.5, a composite picture created by William Notman by cutting out individual negatives of club members and assembling them to record the Montreal Bicycle Club in 1885. The names of the club's officers inset at the top, though very difficult to read, are worth identifying in full. The top row reads: H.S. Tibbs (Committee), J.H. Low (Committee), J.T. Gnaedinger (2nd Lieutenant), H. Joyce (1st Lieutenant), J.D. Miller (Vice-President), C.H. McLeod (President), J.R. Scales (Captain), R.F. Smith (Hon. Sec.), J. Bostell (Committee), A.T. Lane (Committee). The second row reads: G.T. Bishop (Standard Bearer), F.W.S. Crispo (Bugler), W.G. Ross, and G.S. Low. Large numbers of French-Canadians living in the Quebec countryside had begun to migrate to Montreal, and its east end was becoming increasingly francophone.[12] By the 1890s nearly half the population of the city was francophone, and it was a reception centre for a host of other groups, including Jewish, Irish, Italian, and Polish immigrants.[13] Nevertheless, the Anglo-Canadian residents of Montreal's west end formed the city's commercial and financial elite; they dominated Montreal society and occupied most of the key political positions, and the Montreal Bicycle Club remained the preserve of the city's Anglo upper class.

The correspondence published in *Cycling* likewise conveys an elitist picture of Canadian cycling. In December 1890 the term 'knight of the cycle' is used,[14] conjuring an Arthurian image of the cyclist as a noble sportsman defending chivalry on the roads. In the previous issue, an Ottawa correspondent described the higher calling of cyclists; he wrote of their 'purity of motive' and of the 'taste and class of individuals' who participated in 'the noble sport.'[15] A report that the youngest daughter of Timothy Eaton (one of Canada's Methodist merchant princes) had joined the Toronto Bicycle Club in 1890 reinforces the impression that the cycling class was an exclusive group at this time.[16]

Figure 6.5. This is a composite photograph made in William Notman's studio by assembling the negatives of individuals on a background scene. Several of the individual pictures appear elsewhere in this book: for example, Mr Ross (covered with medals at the right) appears in figure 4.4 and the standard bearer (centre rear), in figure 6.7. To use its full title, this is the Montreal Amateur Athletic Association Bicycle Club in 1885. It seems to have operated on more formal lines than some other Canadian clubs did. There are a number of interesting early bicycles in this picture, including the sociable (two-seater) tricycle and the Kangaroo (left), the lever-driven American Star with the small wheel at the front (rear right), and the loop frame tricycle (front right).

In figure 1.10 a revealing image is presented of the social exclusiveness of Canada's bicycle crowd. These cyclists are lined up in front of the imposing summer house of one of the dominion's most senior civil servants, Mr Edward Miall (1838–1903); it was built in Aylmer on the banks of the Ottawa river.[17] The photo was taken at the beginning of the bicycle boom, very likely in 1895; in fact, it is one of at least two photographs taken on the same afternoon, when the cyclists happily lined up and posed, since they wanted to be recorded and counted in on an act of social cohesiveness. Most of their bicycles are new, and the array of outfits and costumes would make a fashion parade. Canada's 'finest' are here on display, and they must have spent substantial sums to become part of the scene.

MEN, WOMEN, AND CYCLING

The social impact of the ordinary bicycle during the 1880s was unequivocally in favour of men; women were effectively excluded from the circle of highwheeling activities, except in the most marginal sense of participating in club social functions.[18] With very few exceptions, riders of the ordinary were men,[19] partly because Victorian dress and behaviour codes made it virtually impossible for a woman to ride one, and partly because the men engaged in the sport constructed bicycling activities as acts of male bonding, often surrounded by militaristic trappings. Many of the photographs from this era have a strong masculinist flavour: the men racing on the Peterborough race track, the Bluenose Tour of 1886, Notman's photographs highlighting the technical aspects of the machines, and the riders taking a breather on Collingwood pier – all these images suggest that the actors were creating a man's world.

The gender relations of the bicycle underwent a dramatic change following the introduction of the safety bicycle. There is abundant evidence that in Canada, as in most other countries, many women took up cycling after 1890. Some idea of their numbers can be gained by examining group photos of Canadian cyclists taken in the 1890s, some of which are presented in this and previous chapters. We have evidence of groups composed only of men, of others only of women, of groups of couples, and of

Figure 6.6. A group of women cyclists pose for P.E.I. photographer A.W. Mitchell on 22 July 1897.

other mixed groups in which men generally are in the majority. In addition, there are a number of pictures of couples out for a private ride.

Three broad conclusions emerge. First, there were significant numbers of women cyclists in all regions of Canada; by the late 1890s they accounted for between one-quarter and one-third of all riders. Second, women bicyclists generally avoided congested urban places and were more likely to ride in parks or the countryside. Third, they might ride in mixed large groups, as couples, or, as shown in figure 6.6, as a group composed exclusively of women.

Women first became active in Canadian cycling in 1890 and 1891. In spring 1891 the correspondent of the Ottawa Bicycle Club reported (in an extraordinary reversal of logic) that several ladies were planning to buy bicycles because 'the prejudices of the ladies are fast becoming dissipated.'[20] Then, on 12 June 1891 the ladies' section of the Toronto Bicycle Club had its initial run, which was a success. A week later one of the women made so bold as to join the men on their weekly ride,[21] and thereafter stronger women riders rode with the men quite regularly.

This development met with barbed criticism in some quarters, but these remarks seem to have had little impact, since the number of women riders in clubs rose steadily after 1891. In fact, in 1895 the Hamilton Bicycle Club became exclusively a touring club, partly as a result of coming out second best in a series of bruising races with the Toronto Bicycle Club, but mainly owing to recruitment of an increasing number of women riders who were more interested in the social side of cycling. In 1897 the Vancouver Bicycle Club reported that it had 205 members, of whom 53 were ladies.[22]

There is no question that many Canadian women took up bicycling in the 1890s, but did this fact represent a significant role change for women? Did it accelerate women's emancipation? Did women achieve an equal voice with men in the cycling world? Did it advance the project of modernity? Within the bicycle clubs, women's adoption of riding made little change to the decision-making process – the club committees and officers remained exclusively male. Patronizing toasts made to 'the lady members' by club presidents at social events testify to the exclusion of women from the power structure of bicycling clubs, where they remained second-class citizens. In this respect, cycling clubs were not much different from most other Victorian social organizations.

If women were active, but marginalized, within the decision-making structures of Canadian cycling clubs, outside the clubs one group became a force to be reckoned with. Under its outspoken and indomitable leader, Frances Willard, the Woman's Christian Temperance Union vigorously promoted cycling to improve the health of women, promote a Christian lifestyle, and lend virtue to modernism. Willard's philosophy was laid out in her best-selling book, *A Wheel within a Wheel*,[23] which has been the subject of a persuasive essay by Phil Mackintosh.[24] Mackintosh interprets Willard's position as one that advocated a particular form of modernity – a responsible and domesticated version that was consciously feminized and in opposition to the masculine modernity of the highwheel era.[25] Willard saw unbridled modernity as a potentially corrupting force that should be tamed and focused by women towards healthy and godly ends. Thus, for Willard and her followers, exercising on a bicycle was both a cure for neurasthenia (a rather unspecific form of nervous exhaustion) and an instrument for women's good health, greater independence, rational dress, and geographical exploration. It also curbed the masculinist excesses that arose when modernity was left in the hands of men. If anything, however, her views reinforced the class divisions built around cycling. In her book (which is dedicated to Lady Henry Somerset!) she describes taking cycling lessons on the terraces of stately Eastnor Castle in England, while in Canada her social connections appear to have been with the wealthier classes.

The remarkable sales achieved by Willard's book, the following it attracted,[26] the rapid growth in purchases of women's bicycles, the numerous advertisements directed to women riders, and the new visibility of the woman bicyclist on streets and in parks all suggest that the bicycle was more than a mirror; it was an instrument for social change, but one whose social impact on women coincided with a series of related changes. What Mackintosh stresses is that, by placing women on a par with men in the most popular recreational activity of the day, the safety bicycle 'taught women how to cope with modernity ... through the bicycle, women could use their moral instincts to learn to be responsibly modern.'[27]

BICYCLE CLUBS AND THE GAY NINETIES

Until around 1891 many American bicycle clubs operated according to a set of rather militaristic rules laid down by Charles E. Pratt in his book, *The American Bicycler*.[28] Pratt proposes a rigid and hierarchical structure for these clubs, one rooted in his military experiences in the American Civil War. Many founding members of U.S. bicycling clubs, such as Colonel Albert Pope, had served as officers during that war and were comfortable with such a system of operation. The first U.S. bicycle club was formed in Boston in February 1878, and Canada's first club was inaugurated on 2 December of the same year in Montreal.[29] The Montreal club, which claims to be the world's oldest bicycle club in continuous existence, operated in the 1880s along quite formal lines, complete with bugler, captain, lieutenants, and standard bearer. By 1882 it had 285 members,[30] and it remained a large club to the end of the century. The fifty-one wheelmen visible in the club's 1885 picture (figure 6.5) represented only a

Figure 6.7. Canada's first bicycle club, the Montreal Bicycle Club, was founded in 1878. Mr G.T. Bishop was the club's standard bearer, a position that reflected the organizational links of highwheeler clubs to the cavalry.

Figure 6.8. Crests of the Montreal Bicycle Club (founded 1878) and the Toronto Bicycle Club (founded 1881), both bearing the Canadian beaver.

fraction of the club membership. The motto of the Montreal Bicycle Club, '*Carpe diem*' ('seize the day,' implying that tomorrow you may not be able to), is shown on its standard bearer's banner in figure 6.7. This motto hints at the fleeting aspects of modernity stressed by Berman and was intended to encourage members to participate in club activities while they still could.

Other clubs dedicated to riding highwheelers were formed in cities in eastern Canada in the years that followed. The Toronto Bicycle Club, for example, was 'organized' in April 1881 and by 1883 had forty members (the crests of the Toronto and Montreal clubs are shown in figure 6.8). Clubs were also founded during the 1880s in Halifax, Saint John (figure 1.3), Ottawa (figure 6.9), Hamilton (figure 2.4), Stratford (figure 3.6), and London; judging by Karl Kron's lists, smaller clubs were operating in the 1880s in several other cities, but the only club west of Ontario at that time was formed in Winnipeg, probably in the mid-1880s. The first bicycle club in the (then) North-West Territories was formed in Regina on 22 May 1892. Although the larger clubs tended to emulate the rather militaristic organization of the American highwheel clubs, the smaller Canadian clubs seem to have operated on a rather more informal basis.

Figure 6.9. Fourteen members of the Ottawa Bicycle Club strike an informal pose in early fall 1886. Smaller Canadian clubs seem to have been less occupied with the militaristic formality of the large U.S. clubs.

In figure 6.9, for example, members of the Ottawa Bicycle Club are shown in a fairly relaxed pose. The umbrella organization, the Canadian Wheelmen's Association (CWA) was formed at a meeting of several Canadian clubs held at St Thomas, Ontario, on 11 September 1882.[31]

The club ethos during the highwheeler phase amounted to a version of Victorian chivalry. Many club members were eligible bachelors who, by parading very visibly on what was a somewhat awesome machine, were consciously acting out the role of 'two-wheeled knights.' Allusions to military chivalry were enhanced by their club uniforms and the parade routines (simple riding patterns look very impressive to onlookers).[32] Since the thrill and danger of riding a high bicycle was widely appreciated, these riders were viewed by their admirers as gallants – rather as some sports heroes are today – but, being drawn mainly from the upper class, they were also assumed to be gentlemen. The records of the Montreal

Bicycle Club show very clearly that marriage was understood to take a rider out of active club life; indeed, members often presented marrying colleagues with what was, in effect, a farewell gift. Thereafter, it was expected that a married man would not attend early morning or evening practices, or races, or regular Saturday rides, but he would attend social events and the occasional ride.

The development of the safety bicycle led to a change in the gender and age composition of the clubs after 1890, but their elite social status remained intact until the end of the decade. The Canadian climate dictated a strongly seasonal pattern of riding, which began in early May. A major outing was usually organized on the Victoria Day long weekend in late May. The Toronto clubs (of which there were eight by April 1892 with between five and six hundred active members)[33] would take a boat across Lake Ontario to Niagara-on-the-Lake and from there ride to Niagara Falls, where they would spend the night. Through the summer the Toronto Bicycle Club (TBC) held evening rides on Tuesdays and Thursdays and a major club run on Saturdays (unless there was a race). No club activities were organized on Sundays: the upper classes were regular churchgoers.[34] This weekly rhythm continued through to September, with Saturday rides maintained up to the Canadian Thanksgiving in mid-October.

The winter activities of the Canadian clubs underlined their social standing. Their main sporting activity was tramps on snowshoes, which they did on two evenings a week and on Saturdays to keep fit. The premises of the (Toronto) Wanderers Bicycle Club (WBC) and the Montreal Bicycle Club included a gymnasium, for those who wanted vigorous workouts, and a billiard room, for more leisurely activity. By mid-November the WBC and the TBC would hold their first 'smokers' (smoking concerts) at their respective club rooms. The TBC had a glee club which would mount a couple of musical evenings each winter and would also hold a 'conversazione,' which consisted of a concert and entertainment followed by a dance in evening dress or club uniform, which ran from 8 p.m. to 2 a.m. The Hamilton Bicycle Club had a 'Mandolin Club,' which organized similar events, including a dance held on Friday 2 February 1894 (see figure 6.10). The twenty dances, ranging from the

Lancers to a Schottische – each listed on a card with a space to enter your dancing partner's name – would amount to a good workout.

The social highlight of the year was a bicycle club's annual dinner, usually held in late winter. On a program surviving from the Hamilton Bicycle Club's annual dinner held in March 1894 a daunting seven-course menu is offered. Annotations on this menu indicate that the diner (Mr A.P. Goering) ate raw oysters, oyster soup, sea salmon with saratoga chips, turkey with cranberry sauce, mashed potatoes, stewed tomatoes and peas, plum pudding with brandy sauce, charlotte russe with ice cream, raisins, coffee, and finally a few radishes to clean his palate. He then listened to a series of six toasts made to the Queen (Victoria, of course), the Hamilton Bicycle Club, the Canadian Wheelmen's Association, the Guests, Sister Clubs, and finally the Ladies. Members of the Wanderers Bicycle Club had written to *Cycling* following their 1892 annual dinner, complaining that the meal had wreaked havoc with their digestive systems, and one imagines that the HBC dinner would have had a similar fallout. Having eaten themselves into what was surely a soporific state, diners then had to endure six speeches ending with toasts. Such a program reflected the class standing of bicycle clubs at that time. Also, despite the participation of large numbers of women in bicycling, the clubs clearly remained very masculinist organizations.

One final indicator of the elite standing, both social and financial, of the bicycle clubs was their assets. In March 1891 it was reported that the membership of the Hamilton Bicycle Club totalled 106, with a cash balance of $500 in the bank.[35] During the next five years the club must have prospered. The program for the Tour of Tours of 1896 (significantly addressed to 'Gentlemen,' not ladies) announces that 'The H.B.C. now owns an hotel at the Falls, and are negotiating for the purchase of one at St Catharines.'

Toronto's cycling clubs were equally big players in the urban real estate market. The TBC club executive had been looking for new premises through most of 1890 and in May 1891 announced that a new clubhouse had been found at 346 Jarvis Street, at that time a prestigious address. A year later the Athenaeum Bicycle Club moved into its new premises,

Figure 6.10. The program of the Hamilton Bicycle Club's 'Mandolin Club' dance of 1894. The musical notation at the top conceals the word 'programme.'

which included 'a magnificent wheel room, capable of accommodating 1,000 wheels' – a huge storage room, by any standards.[36] The eight bicycle clubs in Toronto held a series of discussions on the need for a good bicycle track in Toronto, without which the city could not host the CWA annual meet or similar major events. In June 1892 this problem was solved when a new bicycle racetrack was opened in Rosedale, with a high-quality clay, cinder, and brick dust surface.[37]

METHODISTS, BICYCLES, AND MODERNITY

A rather different set of insights on the complexity of modernity are illustrated by the drawn-out dispute in the 1890s over the operation of Toronto's streetcar services on Sundays and its links to some of CCM's board of directors. The main story of municipal political shenanigans has been told with humour by two leading Canadian historians, Chris Armstrong and Viv Nelles.[38] They make modest claims for the import of their book, entitled *The Revenge of the Methodist Bicycle Company*, yet their story is a compelling vignette of the complex play of modernity and the roles of streetcar operators, bicyclists, politicians, and Methodist entrepreneurs in pursuing different visions of that movement. At the heart of the story is the conflict between those who defended the absolute sanctity of the Protestant Sabbath and proponents of two of the main transport innovations of this period, the streetcar and the bicycle. Its denouement depended upon the different publics served by the two main factions involved.

Armstrong and Nelles's account of how the city's transport infrastructure was modernized can be summarized as follows. In 1891 Toronto was a WASPish city, fully 84 per cent of its inhabitants declaring themselves Protestant and over half belonging to the stricter Methodist, Presbyterian, Baptist, and Congregational sects. About half of the population attended church on Sunday, not only a reflection of the righteous temper of its citizens, but equally indicating there were many citizens with other things to do. City and church politics were closely interwoven. Thus, William Howland was elected mayor in 1886 with strong support from evangelical Protestants and the Orange Order, who gave him a mandate to clean up municipal politics and give some credence to the city's claim to be 'Toronto the Good.' Sunday was to remain a day of reverential silence. The horses that drew Toronto's streetcars rested in their stables: the thirty-year franchise granted to the Toronto Street Railway Company strictly forbade Sunday operation.

In 1888 Howland was replaced as mayor by the more progressive Orangeman, Ned Clarke. Typewriters were introduced to the old city hall on Front Street, the mouth of the River Don was canalized, several miles of sewers were laid in the suburbs, and work began on the new city hall at the top of Bay Street. Bicyclists were delighted when the rough setts and cobblestones of some downtown streets were repaved with a new material known as asphalt, which gave a much smoother ride.

In an 1890 plebiscite, the city voted overwhelmingly to bring streetcars into public ownership in the following year. The majority on Council then put operation of the streetcars out to tender, on condition that the operators electrify the cars and pension off the horses within a year. To recoup some of the costs of electrification, the new franchise operators wanted to get round the 'silent Sundays' that had been in place since 1861; their franchise would be much more lucrative if Sunday revenues were generated by the fixed capital. In 1891 they pressed for a plebiscite on Sunday streetcar operation. It was held on 4 January 1892, and the result was a solid victory for the 'Saints' who garnered 58 per cent of the vote.

By May 1893, with the streetcars now electrified, again demands were made for a plebiscite. The proponents' hope was to hold it in early summer, when citizens could look forward to catching a streetcar on Sunday for a picnic at Toronto Island. Again, there was endless wrangling, and the vote eventually was held on 26 August. It was a closely run race, but the Saints won again with 52 per cent of the votes. Meanwhile, the new franchise agreement signed in 1891 required the company to pay the city $800 for each mile of track electrified, to cover the cost of necessary strengthening of the road bed for the heavier electric cars. The city was obliged to issue a major paving contract to lay permanent tracks surrounded by asphalt – which greatly extended the mileage of smooth roads available to cyclists.

In May 1897 the city voted for the third time in five years on the issue of operating streetcars on Sundays. This time the 'pro' campaign was built on a very broad base, including adherents of all the major religions. Thus, the battle lines were drawn much more clearly, with religion being less of a factor. The case for Sunday streetcars was based, above all else, on the 'changed conditions of *modern* life.' How, its proponents argued, could a large city like Toronto make a claim to be modern if it closed down its major means of transport every Sunday? On polling day, 15 May 1897, there was frenzied activity and the largest turnout in the city's voting history. The final vote was 16,273 for, 16,051 against, with the slimmest of majorities for change. The strongest support for Sunday streetcars came from working men: church-going, middle-class suburbanites generally were opposed to Sunday streetcars. The victors took satisfaction in seeing Toronto 'emancipated from the narrow, mean, domineering spirit of the Methodist Church.'[39]

Did Torontonians rush to ride the Sunday streetcars during that summer of 1897? Not at all! The bicycle boom was now at its peak, and instead of riding the streetcars, many citizens set off on their bicycles to enjoy a Sunday ride. Already by 1895 there were ninety stores selling bicycles in Toronto, plus a number of firms manufacturing and assembling bicycles and parts, spurred on by the 1895 protective tariff. It is hardly surprising, therefore, that there was friction between cyclists and streetcar railway-men. Unruly bicyclists terrified conductors, while 'particularly upsetting to riders was the streetcar company's habit of watering its tracks during the morning and evening rush hours, making the pavement as slick as ice.'[40]

When they started operating at the peak of the bicycle boom, Sunday streetcars had few riders. The entrepreneurs who had lobbied, bribed, and manipulated for the 'yes' vote initially found Sunday operation an unprofitable exercise. Yet there was a delicious irony to this outcome: among the stalwart Methodists in opposition to Sunday streetcars were the proprietors of the main bicycle companies, including the Massey, the Harris, and the Flavelle families. At least for a few years the Methodist Saints could relish the low numbers of streetcar riders. This revenge was short lived, however, since by 1900 the bicycle craze had fizzled out and CCM was in financial difficulties. The Toronto Railway Company, by contrast, went on to make very substantial profits for its shareholders for the next twenty years.

This episode also captures some of the conflicting currents of modernity. One group of modernists stood firmly for technological innovation, their gaze fixed on future possibilities. Ranged against such forces, however, were the likes of Sir Joseph Flavelle, whose goal was to apply Methodist doctrines of thrift, orderliness, and accountability to the workplace. Flavelle looked back for inspiration to John Wesley, who a century earlier had laid out his vision of Enlightened Christianity in *Deed of*

Declaration: Wesley sought to replace the mysteries of Catholicism with a rational form of life based on applying modern methods to all aspects of Christian life.[41] Flavelle's biographer, Michael Bliss, suggests that he became modern by respecting the past, specifically the recent past, when Enlightened thinking had permeated western thought.[42] Flavelle's business methods, which included careful accounting systems, concern for the welfare of employees, delegation of responsibility to managers, vertical integration, and product innovation, grew from his respect for Methodist philosophies that had been worked out during the Industrial Revolution a century earlier. This vision of modernity contrasted with that of the technological innovators of the late 1890s, who paid little attention to earlier prescriptions.

THE BICYCLE AND SOCIAL MODERNITY

What conclusions can be reached about the impact of the bicycle on social modernity? In her critique, Anita Rush debunks a number of myths associated with the bicycle boom, four of which have social implications that help to focus this summation.[43] First, Rush contends that the bicycle did not erase social barriers. The evidence presented above indicates that this is correct, but it is also entirely consistent with the process of modernity, which typically (but not exclusively) is led by a group of social innovators. By the time the mass of consumers have caught up with any particular fashion, innovators must have adopted new fashions and new cultural forms if the modernity bandwagon is to keep rolling.

Second, Rush is of the opinion that, contrary to popular belief, cycling had a minor impact on women's clothing fashions and, more generally, on the informality of everyday affairs. This is correct in the sense that a host of social changes affecting women were under way during the 'gay nineties.'[44] In ballet, for instance, the hemlines of women dancers rose above the ankle at the same time as women cyclists began to wear shorter skirts.[45] Other sports, such as croquet, golf, and tennis, and other innovations, such as electricity and the typewriter, formed part of this broad social transformation. Again, this pattern is entirely consistent. The process of modernizing daily life never hinged on one activity, but relied on the continual diffusion of a variety of new artefacts, new ideas, and new attitudes, which later would be discarded and replaced by yet another set of new tastes and fashions as the new became old. Yet it is not correct to suggest that the bicycle was simply a 'mirror,' a passive reflector of its surroundings. It was an active instrument – one among many – in the modernization of gender relations in late Victorian society.

The third myth rejected by Rush states that the bicycle produced a new breed of tourists who demanded road maps, better country inns, and improved roads to ride on. She argues that, in practice, the bicycle-touring phenomenon was only one manifestation of a social trend towards increased leisure travel. It is shown in this chapter that bicyclists were active tourists, and their clubs became hotel proprietors. Moreover, many hotels and inns advertised regularly in cycling magazines. In addition, a number of bicycle maps were drafted, including the one illustrated in figure 5.9. Although tourism (for instance, to the Rockies by train and up the Saguenay River by boat) was growing, the bicycle is seen to be especially important at the local level in creating a new geography of travel and movement. So, although Rush is correct in identifying a general trend to increased leisure travel, the bicycle is seen to have made a unique contribution. This issue is taken up in more detail in the following chapter.

The final argument made by Rush is that cyclists had little social impact, owing to their sheer lack of numbers. She cites, in evidence, the absence of cyclists on Sparks Street in Ottawa (a curious choice, since most cyclists preferred riding in the countryside) and the fact that photos featuring several cyclists (such as figure 1.10) are posed photographs and therefore provide unreliable evidence. I do not accept this argument. There are, of course, many images taken in Canada in the 1890s containing no bicycles; bicycling was a seasonal activity, and there were only five months when it could be done comfortably. Yet, in season, substantial numbers engaged in cycling. *Cycling* reports, I believe quite reliably, that in 1891 Toronto had close to 600 club members and 1,500 wheelmen (implying that there were around 900 riders who did not belong to clubs); these totals grew through the rest of the decade. Rush notes that in 1897 Calgary had only 120 cyclists in a population of 4,000 (she views this as a low number); as explained in endnote 34 of chapter 5, these figures probably imply that about one in three middle-class men aged twenty to forty were bicyclists! Figure 6.11, a photo taken in the late 1890s in Stanley Park, Vancouver, hardly suggests that bicycling was a pastime for a small minority. In addition, Rush's own observation that 395 cyclists passed through the junction of Yonge and King streets in Toronto during a thirty-minute interval early on a summer evening in 1895 (the junction shown in figure 1.14) reflects a high level of bicycle use. Moreover, the unposed image of Sparks Street shown in figure 5.11 (with six bicycles in view) somewhat undermines Rush's own particular case study.[46] There is considerable evidence, therefore, that, in season, bicycling was a popular activity during the 1890s, but in the context of modernity, there is more to be said on the question of numbers. If almost everybody rode bicycles, as is true in China today, then bicycling would be a mass activity with no social cachet. To the extent that one stream of modernity was associated with the consumption of positional goods and services, available in limited supply, by a group of elite trendsetters, then bicycling need not be massively popular to serve as a minor carrier wave. Until the final years of the decade, bicycling was popular among only well-off Canadians, and when the masses did take to cycling, the elite moved on to various new pursuits.

In sum, what were the social manifestations of the bicycle? One explanation leads the discussion back to the issue of modernity. The chronological age of bicyclists broadened dramatically with the advent of the safety bicycle, and by the end of the 1890s in some cases it had become a family activity. Women joined the ranks of Canadian bicyclists in large numbers after 1891, but the core of Canada's bicycling crowd remained members of the Anglo elite until the end of the bicycle boom. One of the major mechanisms driving the project of modernity was the succession of technological innovations that were incorporated into production systems and adopted by social and economic trendsetters, usually people who could both afford the new objects and who wanted to be seen adopting them. These business and professional people had a direct interest in promoting economic innovation. Note that bicycle prices remained high until the late 1890s, partly because the succession of bicycle-related innovations allowed manufacturers continually to reap new technology rents,

Figure 6.11. A bicycle jam in Stanley Park, Vancouver, circa 1898. There are nearly 100 bicycles in this photograph, which may be of a ride organized by the Terminal City Cycling Club.

that is, premiums paid for a technologically superior product that was in limited supply.[47] Patents were often used to sustain this system and to stop cheap imitations from flooding the market. Thus, the bicycle crowd expanded to include women and both younger and older riders, but they were still mainly members of privileged classes with a collective interest in the production and consumption of bicycles.

It became clear, by the late 1890s, that the bicycle industry was no longer making many significant technological breakthroughs. New models were announced each year, but the changes had diminishing substance, and this fact eventually punched a hole in the fabric of planned obsolescence. If a second-hand 1897 bicycle was almost indistinguishable from a new 1899 bicycle, why bother to trade up? In 1898 the Pope Manufacturing Company introduced the chainless (shaft-drive) bicycle to loud fanfare in the hope of prolonging the bicycle boom, and Canadian bicycle manufacturers – Massey-Harris, Goold, Cleveland, and Gendron – rapidly followed suit. It soon became clear, however, that the shaft–driven bicycles were less efficient, and their sales tailed off. In practice, bicycles manufactured fifty years later were not fundamentally different from those made in 1898; they became commonplace, and they no longer had a social or technological cachet. Trendsetters, therefore, began to look to other avenues of consumption into which the energies of modernizers could be channelled. Electric lighting, gramophone, radio and telephone systems, automobiles, Bakelite products, and aeroplanes were among the promising possibilities on the horizon.

As bicycle-related innovation decelerated, the possibility of mass production of bicycles increased. The bicycle entered the mature stage of the product cycle.[48] Mergers occurred, cartels were formed, and the price of a 'popular' mass-produced bicycle dropped to as low as $30. The bicycle could no longer offer the social elite a means of defining itself once the clerical and labouring classes began to join the ranks of wheeled citizens. Thus, in Canada, the most important social relations of the bicycle were not with age or gender, but with class. This argument appears to be borne out by the history of Canada's bicycle clubs in the 1880s and 1890s and the dispute over the operation of Sunday streetcars in Toronto.

The social impact of the bicycle was therefore initially confined to the middle classes. Working men, their wives, and their children may have dreamed of owning a bicycle, but until the last year or two of the era, the price tag remained beyond their reach. The growth of bicycling through most of the 1890s was achieved largely by attracting both younger and older riders from the same well-to-do class that rode highwheelers and by bringing women into the socially restricted world of bicycling. In both cases this was made technically possible by the availability of light and reliable safety bicycles. The increasing number of women cyclists in Canada, however, did not herald a significant change in gender relations; Canada's bicycle clubs remained as masculinist as ever. The bicycle clubs remained socially exclusive; indeed, the cost of club activities alone would exclude the working man even if he could afford to purchase a second-hand machine. Besides, club events did not take place on Sundays, when he had his only day off work. Yet for those who bicycled, there were new places to be experienced and a new understanding of space to be learned. The bicycle brought in its wake a new geography.

CHAPTER SEVEN

Larger Spaces and Visible Places

We have seen that in the latter part of the nineteenth century the bicycle constituted a carrier wave. Although relatively minor in comparison with some of the great innovations of history, the bicycle nevertheless was of considerable importance both in advancing the project of modernity and in providing a crucial bridge between the railway age and the era of the automobile. It triggered a burst of bicycle-related innovations, it changed patterns of industrial production, it boosted new forms of consumption, and it allowed the bicycling class to show off its colours as a social elite. From a geographical perspective, these elements formed a nexus by producing a new geography in which individual action spaces began to expand and in which new kinds of visible bicycling places came into fashion. New geographies were therefore produced by the ride to modernity. In this chapter we will examine these new geographies, first, by considering the ways that bicycling transformed space, and second, by exploring the relations between the identity of the cyclist and the particular places that cyclists sought out to establish and solidify that identity.

LARGER SPACES 1: THE WORKING BICYCLE

Canadians have never made use of the bicycle as a working machine to the same degree that most Europeans have. Owing partly to the Canadian climate, partly to the great distances separating major towns, and partly to a culture that has been enamoured with the automobile since early in the twentieth century, the bicycle has had a more limited practical use in Canada than in countries like the Netherlands, France, and Denmark. Since the early days of cycling, few Canadians have taken advantage of the machine's potential either as a means of travelling to and from work or to perform their jobs. Nevertheless, some have grasped its possibilities.

Two of the earliest recorded cases of the bicycle's being put to practical use in Canada involved physicians. It was noted in chapter 1 that in March 1869, or soon thereafter, Dr Robertson of Stratford, Ontario, purchased a boneshaker on which to make house calls. Another early example concerns Eli Franklin Irwin, who was born in 1867, the year of Confederation. At the age of nineteen, Irwin gained admission to the University of Toronto School of Medicine, graduating four years later with an MD degree. Irwin often spent the weekend at his mother's house in Newmarket, some twenty-seven miles from his lodgings in downtown Toronto, periodically making the three-hour trip home up Yonge Street on his highwheeler. While at home he would stock up on food and provisions for the next week, carrying them in a back pack when he made the return journey to Toronto on his bicycle.[1] These examples, however, appear to be isolated cases; in Canada the boneshaker and the ordinary bicycle were used largely for pleasure and sport.

The safety bicycle, by contrast, did achieve some popularity as a practical means of delivering goods and messages and as a vehicle for commuting to work. In figure 7.1 a group of about twenty teenage messenger boys employed in 1900 by the GNW Telegraph Company is depicted. Supervised by some stern-looking elderly gentlemen, these youngsters had the task of delivering and collecting cables across downtown Montreal on bicycles. Unlike the mountain bicycle messenger services operating in Canadian central cities today, these young men did not have to keep a wary eye out for aggressive automobile drivers, but considerable care was needed when they crossed tram lines and passed horse-drawn vehicles. Judging by the warped rims and bent handlebars visible in this photograph, several of the bicycles ridden by these messenger boys had seen front-line action on the streets.

Figure 7.1. A group of young messenger boys pose on the steps of the GNW Telegraph Company office in Montreal, 1900. Presumably, these boys worked on foot when the winter snows fell.

The use of the bicycle to commute to and from work involved similar hazards. One of the more notorious early cases of road rage concerns. Mr Richardson, a stalwart of the Toronto Bicycle Club, who was obstructed while pedalling to work. By 1891 Richardson had been riding to work in the downtown area for two to three years.[2] On a number of occasions he met a 'smart aleck' driver, who habitually turned his horse to block his way, forcing him to dismount. Richardson eventually found out the driver's name and employer and took legal action against him. The driver, Mr Powers, was brought before a court, where he was convicted of obstructing other vehicles on the road and required to apologize to Richardson and pay his costs. The judge remarked in his ruling: 'You thought the bicyclist had no rights on the road, did you? When this court has done with you, you will know differently. I suppose you have no idea that this young man's life is not worth that of your horse.'[3] Incidents like this suggest that it took some time for bicyclists to negotiate their rights of way with long-established road users. The new geographies made possible by the bicycle were not constructed overnight.

An enterprising example of a bicycle's being used for work concerns a seamstress, Susanna Frances, born in 1866, who plied her trade around rural southwest Ontario in the 1890s (see figure 7.2). Miss Frances would pedal her bicycle to a farm or house, where she would take up residence and make one or more dresses for the women living in the household. A wedding, or a daughter's achieving an age where marriage became a serious prospect, might lead to such a visit. Miss Frances would stay with a family for several days and possibly for a few weeks, while she helped her client to pick suitable fabrics and then sewed the required dresses. Her hope was that by the time one assignment was complete, another commission would be secured. In rural Canada, seamstresses and milliners had led this kind of peripatetic existence for some time, using trains and buggies as transport. What was audaciously modern about Susanna Frances was that she travelled independently by bicycle from place to place.

The most common practical use of the bicycle in the 1890s was made by workmen travelling from their homes to their workplaces and back. Bicycle riders were exposed to the elements, of course, which in Canada could be quite extreme, but a bicycle could move at about the same overall

Figure 7.2. Miss Susanna Frances, a travelling seamstress who pedalled from customer to customer in Southern Ontario in the 1890s. This photo was taken circa 1898 in Carrick Township, Bruce County, Ontario.

speed as a streetcar (since it didn't have to stop frequently), it travelled door to door, and it cost less to maintain than a daily streetcar fare. It seems likely, for instance, that some of the 395 cyclists noted as having passed through Toronto's Yonge-King intersection between 6:00 and 6:30 on a summer evening in 1895 were heading home after work. In 1892 it was reported in *Cycling* that Toronto's new Confederation Life Building was equipped with a bicycle storage room in its basement for tenants and employees who rode to work.[4] There are reports that during the bicycle boom of the mid-1990s some of Ontario's civic leaders cycled to work. Sir James Pliny Whitney, who later became premier of Ontario, cycled regularly between his home and the new legislative buildings at Queen's Park, always sporting a bowler hat (a derby).[5] By the turn of the century, as the cost of bicycles began to decline, working men also began to bicycle to work. By this date, working bicycles were not ostentatiously displayed, because they had been demoted from a status symbol to a functional form of transport.

One other practical use for the bicycle, which combined the machine's functional with its display roles, appeared towards the end of the era. The military possibilities of the bicycle had been promoted by men such as Colonel Pope since the 1880s. During the First World War several divisions of cyclists were formed by the Canadian army, although I have found no record of the Canadian Army's formally adopting the bicycle for military purposes before 1900. On the other hand, as the advertisement for Massey-Harris bicycles shown in figure 3.15 reveals, Canadian-made bicycles were adopted by Australia's Queensland Imperial Bushmen; since these machines saw service in South Africa, it can be legitimately claimed that Canadian bicycles were put to military use during the period under consideration.

LARGER SPACES 2: THE KLONDIKE

The most peculiarly Canadian use of the bicycle, one that has developed a unique mythology, most surely occurred in the Klondike. Following the discovery of gold in 1896, a wave of migrants headed north to the Yukon, lured by dreams of striking it rich. Here, the bicycle supposedly was put to good use as a means of transport and as a kind of two-wheeled barrow, which could be loaded with possessions and pushed over the packed snow. The story is told with verve by Pierre Berton in chapter 6 of his book on the Klondike gold rush, entitled 'Balloons, boatsleds and bicycles':[6]

> The so-called 'Klondike Bicycle' was a popular item with the stampeders, for the Klondike strike came at the height of the bicycle craze ... Such was the faith in the bicycle that thousands were prepared to believe that this was the ideal way of crossing the mountain passes [to the Klondike]. Two youths, cycling around the world for a Chicago newspaper, switched

> plans and began propelling themselves towards Alaska. On September 20 the Misses Olga McKenna and Nellie Ritchie, described as 'two of the best wheel-women in Boston,' started pedalling north, announcing that they expected to enlist one thousand women in their move to cycle to Dawson City.
>
> A New York syndicate was meanwhile busily marketing machines designed especially for the stampede to which were attached four-wheeled trailers with a freight capacity of five hundred pounds ... Two New Yorkers in the fall of '97 left for the Klondike on a strange contraption consisting of two bicycles joined together with iron bars heavy enough to support a small rowboat containing their outfit. They declared they would reach the Klondike in ninety days by this method, but the ensuing winter found them still at the foot of the White Pass.
>
> ... another syndicate of wealthy New Yorkers was reported engaged in building a bicycle path to the Klondike to service a chain of trading posts.

Here, the mystery begins. For in the illustrations in Berton's history of the Klondike gold rush and his subsequent photographic essay on the same subject[7] there is not a shadow of a bicycle, except for figure 7.3, in which is shown an outfitter's shop in Vancouver in 1898 selling would-be miners all the equipment they might need (and some they did not need). Moreover, this illustration is ambiguous: the two bicycles are presumably being sold as part of the Klondike outfit, but it is possible that two young bicycle owners were simply showing off their latest acquisition.[8] If bicycles were so much a part of the Klondike rush, surely they would appear in street scenes of Dawson City.

Where were all the Klondike bicycles? Enquiries made at the photographic archives in Dawson City and Whitehorse did unearth a few images that demonstrate that the bicycle was used as a working vehicle in the north. In figure 7.4 a bicycle stands outside the Hotel Thistle, roadhouse number 79 on the 400-mile stagecoach line from Whitehorse to Dawson City. These roadhouses were spaced about every twenty miles to service travellers. An 1899 studio portrait of B.H. Svendson, who rode a bicycle on the stage line from Skagway in Alaska to Dawson City via Whitehorse, was taken to preserve a record of his bicycle journey made

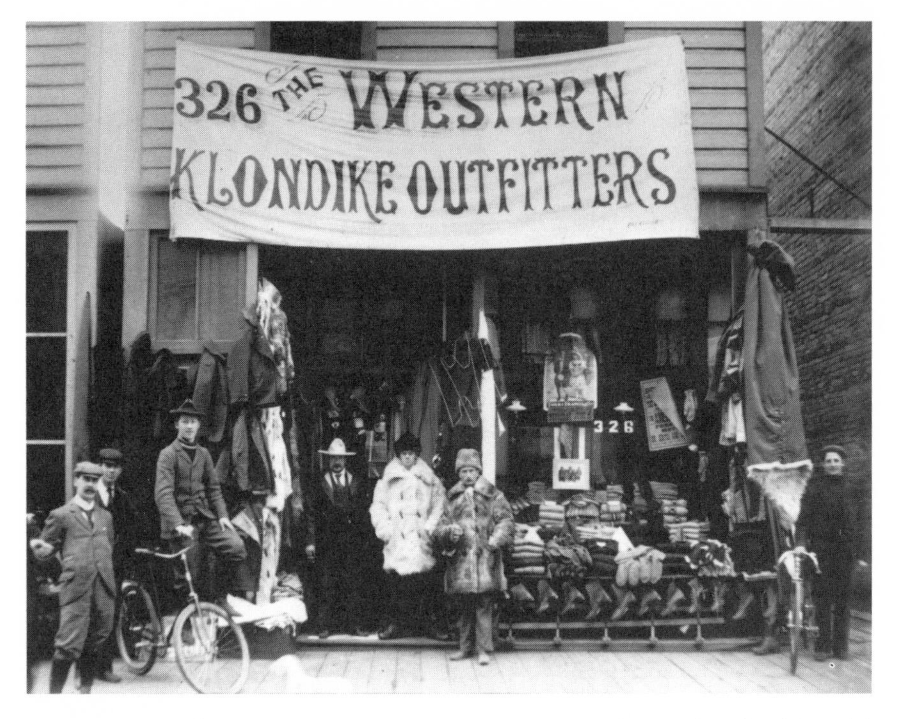

Figure 7.3. Cordova Street, Vancouver, May–June 1898, during the Klondike gold rush. Stampeders had little money left by the time they had bought their boat ticket and mining outfit. Many of the economic benefits of the Klondike flowed rapidly back to Vancouver.

under adverse conditions. The photographer obligingly recreated these conditions in his studio.

Bicycles were usable in the Yukon in winter on packed snow and on ice and in summer when the ground dried out, but not during spring and fall, when slush and mud made the trails and lakes impassable. William Humber provides some graphic details of riding in the far north.[9] The rivers, lakes, and swamps that account for a large part of the surface area of the Yukon presented the biggest hazard. If a bicycle broke through the ice surface a rider would land in the water; having extricated himself, he would then have to kick off the ice that rapidly formed on his chain and wheels before being able to continue. Smooth ice surfaces on a lake or

Figure 7.4. Roadhouse number 79, on the Whitehorse–Dawson City trail towards the end of the gold rush (circa 1900). A stagecoach loaded with passengers and freight stops at the roadhouse, Hotel Thistle. Horses and, in winter, dogs were the principal means of transport in the Klondike, but a few bicycles made the journey in the latter part of the gold rush, including the one illustrated here.

river presented so little friction that a rider might slither in any direction, while a strong tailwind could blow a bicycle, even with the brakes on, across a lake at such high speeds that the tires and brakes would then need cooling in a snow bank. Under such conditions, progress into a headwind was impossible.

Cold weather notwithstanding, bicycles were a practical means of transport for much of the year. The photograph in figure 7.5, taken on 20 October 1899, shows what were probably the last two scows arriving in Dawson City that season. The bicycle may well have belonged to a local merchant, Mr Coppins, who is checking that his supplies have arrived safely.

Figure 7.5. Two scows loaded with 32 tons of moose meat arrive in Dawson City from Bennett, B.C., in October 1899. The substantial amount of snow lying on the ground did not stop cycling in the Yukon.

Perhaps the most extraordinary case of the bicycle's being put to use in the north concerns the Reverend Mr Wright, an evangelist, who pedalled his Massey-Harris bicycle over packed snow from one miners' bunkhouse to another, spreading the gospel (figure 7.6). His photo was taken outside Roadhouse 85, a log cabin in which the photographers, Lars and Duclos, had set up shop. The temperature is reported to be -40°F.

We can conclude from the above evidence that bicycles were not important during the early years of the Klondike, when stampeders had to manually lift all their possessions up the massive ice wall of the Chilcoot Pass, but when the slightly easier access route via White's Pass was established, a number of machines were brought in. The complete absence of bicycles from every available street photograph, however, suggests that their importance may have been exaggerated by some historians.[10] On the other hand, the Klondike did sufficiently capture the public's imagination

Figure 7.6. The Reverend Mr Wright on an evangelical tour of the Klondike on his Massey-Harris bicycle, probably in 1899. On the hard-packed snow of a Yukon winter, the bicycle was a practical means of travel.

that one manufacturer, the Mead Cycle Company of Chicago, named two of its 1898 models 'The Klondike' and the 'Ladies' Klondike' (both priced at $40).

What are the implications for the larger issue of modernity? The fact that Svendson and the Reverend Mr Wright had themselves photographed specifically with their wheels suggests that their bicycles were important to the message they wanted to convey. They had stretched space by travelling to new frontiers on their bicycles. They were creating a new geography in which travel to the ends of the earth was increasingly the norm. Travel to distant places by the latest machine identified these frontiersmen with the project of modernity.

LARGER SPACES 3: SHRINKING DISTANCES

The single most important impact of the bicycle, in Canada and elsewhere, was to allow bicyclists to travel over land distances that had previously been unthinkable. Bicycles were not like trains and streetcars. Riders did not have to follow the rails to a fixed schedule, stopping at fixed points. On a bicycle they could travel along almost any right of way, stopping and turning when they wanted, in ones or twos or larger groups. A bicycle was even more under the control of its rider than a horse, it did not tire, nor did it need water and fodder, although it did sometimes have a puncture. It was, in short, a geographically liberating machine, serving as a herald for even greater liberation in the age of the motorcar and aeroplane.

This geographical liberation extended the pulse of modernity into previously marginal regions in a manner already described by Rob Shields in his book, *Places on the Margin: Alternative Geographies of Modernity*.[11] The race for modernity has always been geographically focused, mainly on larger cities, while smaller places in marginal regions are continually left behind. Movement, in the form of transport, communications, and migration, has the potential to connect the (modern) centre with margins occupied by people so far unaffected by the latest waves of progress. Thus, on afternoon rides into the countryside, on excursions further afield by train followed by a longer ride home (see figure 1.5), on weekend bicycling trips, and, for the most adventurous, on expeditions to foreign lands, bicyclists were consciously parading their modern lifestyles before others occupying the geographical margins. People living in the margins had two main responses. Some resented this intrusion of popular culture into their world and in small ways resisted, through actions such as putting sticks through spokes, obstructing the way with animals and horse-drawn vehicles, and being unwelcoming. Others looked on the new artefacts with envy, sometimes with awe, hoping that they, too, or at least their children, might eventually join in. Indeed, by 1900 their hopes began to materialize.

With new bicycles costing as little as $30 and second-hand ones selling at half that amount, those who were economically marginalized could join the growing flotilla of cyclists. Alas, they were chasing a moving target that had already shifted. By 1900 new patterns of consumption and new popular forms of leisure allowed those in the vanguard of modernity to connect to a new centre with even remoter geographical margins. This process, whereby progress at the centre is slowly spread to the social, economic, and geographical margins, operates to this day as modernity continues its endless cycle of renewal and disposal.

The actions of cyclists during the bicycle era reveal that this process of diffusion was set in a hierarchy of centres and margins. In Canada, records survive of bicycle excursions being made on local, regional, foreign, and world scales.

The vast majority of bicycle rides were local trips, on which cyclists travelled ten or twenty miles out of the city into the surrounding countryside. Many of the illustrations presented in previous chapters were taken on just such outings. The most popular destination within Canada was probably the Half Way House (figure 7.7). Beloved by a generation of Toronto cyclists, it was conveniently located fifteen miles east of the city on the Kingston Road at Highland Creek – the perfect place for refreshment before the ride home.[12] Two other photographs capture this idea of cyclists' exporting modernity to the local margins. In figure 7.8 three Montrealers out for a ride along the south shore of the St Lawrence River are shown. Taken about 1900, the photo speaks to a dialectical relation between the traditional and the modern; it juxtaposes traditional rural elements – the horse, the boardwalk, the country house built in the Quebec vernacular style, and the dirt road – with the three cyclists parading new wheeling technology, their innovative clothing, the new reality of city dwellers pedalling for pleasure through the countryside, and the photographer who captured the image.

As new lifestyles spread to smaller towns, they, in turn, spread new fashions to the surrounding countryside. Figure 7.9 is an engaging photograph taken in Tavistock, Ontario, in about 1896, of a young couple about to set off for a ride on local country roads on a sunny summer afternoon. Tavistock was one of those small country towns that followed,

Figure 7.7. The Half Way House was built on the Kingston Road at Highland Creek, in 1849. It marked the halfway stopping point on the stagecoach line to Dunbarton (near Pickering). Forty years later, the inn became the favourite destination of Toronto's bicyclists. This photo shows fourteen bicycles, most with the sloping top tubes and quite large inflatable tires that were in favour in 1894.

Figure 7.8. An outdoor cycling scene on the South Shore of the St Lawrence River, circa 1900.

Figure 7.9. This photograph (circa 1895), printed from a glass negative, lay behind a panelled wall in Tavistock, Ontario, for about fifty years until it was discovered, along with hundreds of other photos, in the 1980s. Neither bicycle has a brake; riders could stop by pushing against the pedals, which slowed the rotation of the chain and back wheel.

perhaps by a year or so, the fashions of larger centres, such as Stratford or Woodstock, but that then passed them on to the neighbouring villages. For three decades the local photographer, Karl Lemp, recorded people and events of his home town with a compassionate eye. He photographed the floods and blizzards that nature visited upon its citizens, but his favourite theme was the chronicling of its material progress. As both a photographer and a bicyclist, Lemp himself played an active role in introducing modern ways to this country town. He has an instinctive feel for the freedom this couple has achieved though their bicycles. Unencumbered by any luggage, and dressed quite casually, they are about to set off as their fancy takes them.

Figure 7.10. Around 1900, a young lady takes a bicycle ride on a summer day to visit a friend living in the Ottawa Valley. Note her skirt and bicycle cap and the chain guards on the bicycle.

A similar sense of changed rural life is evident in figure 7.10. The setting is a farm in the Ottawa Valley on a pleasant summer day in the last year or two of the century. A teenage girl has pedalled over to visit a friend on a neighbouring farm. As she is evidently thirsty from her ride, the friend pours her a glass of lemonade. At first glance, this is hardly a riveting story, but in its particular context, the event illustrates the new freedom to travel around the countryside made possible by the bicycle. Nearly a century later, the snowmobile had a similar liberating effect by allowing snowbound residents of the Canadian north to travel around in midwinter, when previously they had been housebound. Further evidence of changes in rural life are to be found in figure 7.11, where members of a farm family

Figure 7.11. A family picnic at Arran Lake in Bruce County, Dominion Day, 1 July 1898.

living in a distant part of the Queen's Bush, Carrick Township, in County Bruce, Ontario, are shown on their annual Dominion Day picnic at Arran Lake. The year is 1898; the bicycle is now making an appearance at the geographical margin. Note that the parents and the old folks travelled on the horse-drawn cart; it was the youngsters who pedalled.

LARGER SPACES 4: THE TOUR OF TOURS, 1896

Since the 1880s bicyclists had been regularly pedalling their high bicycles and safety bicycles into the countryside surrounding Canada's major cities. A few groups had embarked on organized tours, one of the earliest being the Elwell Bluenose Tour made by a group from Portland, Maine, to New Brunswick in 1886.[13] By the mid-1890s rather more adventurous trips were being organized by bicycle clubs. By serendipitously matching up a set of glass slides in the National Photo Archives with a tour program,

Figure 7.12. These riders are evidently enjoying the 1896 Tour of Tours organized by the Hamilton Bicycle Club for the May long weekend.

it has been possible to reconstruct one such weekend trip quite accurately. The Hamilton Bicycle Club took the opportunity presented by the long-weekend holiday on the 23, 24, and 25 May 1896 to organize a ride to Niagara Falls. Fortunately, there was a photographer in the group; indeed, three of his images have been used to illustrate previous chapters (figures 2.6, 5.3, and 5.4). Cyclists on the Tour of Tours set out after lunch on Saturday, riding to St Catharines, where they spent the night at the hotel shown in figure 5.3. Sunday's ride took the group to Niagara Falls, where the club had recently purchased another hotel. All day Monday was available for the ride back to Hamilton. It was obvious that much fun was had on the way, including the 'this way! no that way!' pose by the tandem pair in figure 7.12. This illustration also points to another important transformation of the countryside, which was becoming more 'Cartesian': cyclists have here signposted the countryside (close inspection shows the letters CWA – Canadian Wheelmen's Association – beside the Hamilton sign).

For city residents the countryside was becoming a recreational space, but this change required the posting of direction markers to structure that space for its new users.

Similar tours were organized by most of the larger cycle clubs in Montreal, Toronto, and elsewhere. Their significance lies not only in the novel idea of a club group taking off to tour the countryside in a fairly leisurely fashion, but also for the way they inaugurated a new kind of holiday. These trips were precursors of the modern package holiday, such as sun or ski breaks or trips to Disney World. To this point, tourism had involved highly structured events, such as railway journeys to spas such as Banff Springs; transhumance to cottages on Lakes Muskoka or Mephramagog for the duration of July and August; and cruises on the Great Lakes or the St Lawrence and Saguenay rivers. With their flexible programs and light agendas, these bicycle tours broke new ground that tourists built upon in the twentieth century as motor transport stretched space even further to create new geographies.

LARGER SPACES 5: A CYCLE TOUR IN EUROPE

A handful of Canadian cyclists had the time, the resources, and the determination to venture even further afield. Notable among them was Constance Rudyard Boulton, who in 1895 sailed to Europe with a female travelling companion for a cycling vacation and on her return reported the adventure in five rather verbose chapters published in the *Canadian Magazine*.[14] The first chapter ends with the arrival of their ship in Gibraltar, their bicycles remaining stowed away in the ship's hold. Two land visits are made in the following chapter, to Gibraltar and Algiers, but in neither case was it possible to transport the bicycles to shore, nor could a 'guide on a wheel' be found. In the next episode the travellers arrive in Naples, where their bicycles, along with the machine of a fellow traveller from the United States, were impounded by Italian Customs for twenty-four hours. Having negotiated the payment of a customs indemnity, they were released; the women then took several spins along the waterfront and visited various museums and archeological sights to admiring cries of 'Bravo Segnorena [*sic*].' Indeed, the two women cyclists caused a major stir wherever they went. Here, at the margins of nineteenth-century industrial Europe, women cyclists were a thoroughly modern spectacle.

Vesuvius was tackled next, on foot and by donkey, followed by a very rough, three-hour ride on a Roman road to Pompeii. After touring the buried city, the cyclists collected their bicycles, only to discover that one had a puncture, which forced them to return to Naples by train, where an enterprising bicycle shop owner charged an exorbitant fee to fix the tire. After several more rides in the Naples area, the two adventurers took the train to Rome, where they spent seven weeks riding the streets without bells, lamps, brakes, or licences, in each instance contravening Italian law. There followed a train trip to Florence, where again they made daily trips on their bicycles. They also sought out Italian women cyclists, with a view to comparing styles, coming to the following self-congratulatory conclusion: 'we rode to the beautiful Casciné to see what Italian women looked like as cyclists. With low, stumpily-adjusted saddles, ill-arranged skirts, and flower-bedecked hats, they did not look trim, and after seeing numbers of women of different nations riding, I have come to the conclusion that Canadian girls take the palm every time.'[15] Taking a train from Florence to Bologna, the travellers then pedalled for three days via Ferrara, Montselice, and Padua to Venice, to complete their Italian journey.

Such a cycling holiday was a novelty from both a Canadian and an Italian perspective. The presence of two unaccompanied young women cycling in fairly casual dress was a wholly new phenomenon in Italy, which explains why they were thronged by crowds on several occasions. The fact that the editor of the *Canadian Magazine* ran a lengthy serialization of their story indicates Canadian interest, particularly in the way the bicycle broke geographical barriers. The leitmotif of the series was the spatial freedom enjoyed by bicycle riders.

LARGER SPACES 6: KARL CREELMAN'S WORLD TOUR

The potential of the bicycle to shrink the globe was taken to its limit by Karl M. Creelman of Truro, Nova Scotia, who in 1899 said farewell to his family and friends and rode westwards out of his home town with a goal of circumpedalling the globe. Brian Kinsman, in his excellent memoir on

Creelman's adventure, shows that the whole enterprise was quite improbable in several respects.[16] First, Creelman was a not a stereotypical hero: he stood five feet six inches tall, weighed 130 pounds, and was a rather shy twenty-one year old (figure 7.13). Second, he was of modest means; his father was a carpenter and Karl worked in Truro in the stationery business. Third, although the American cyclist, Thomas Stevens, had achieved renown pedalling his highwheel bicycle some 13,500 miles around the world between 1884 and 1887, a number of riders subsequently attempting this journey had met with disaster. In 1892 an American bicyclist had been murdered by Turkish tribesmen, while in the previous year two Americans from the midwest had had to shoot their way, frontier-style, out of a tight corner in China. Fourth, he did not know how to ride a bicycle and he did not own one; moreover, his efforts to persuade a Canadian manufacturer to donate a bicycle for this trip had met with failure. Eventually he purchased a second-hand New Barnes bicycle in Truro, took a bicycling lesson, and set off on 11 May 1899.

By July he had reached Brantford in Ontario, where his first bicycle gave up the ghost. However, there he did persuade the Goold Bicycle Company to give him a Red Bird Special in return for promoting their machines on his journey. He then rode west to Chicago and north to Winnipeg to cross the prairies, using the railway bed as a road much of the way. His western Canadian adventures with a gang of murderous tramps, an attack by a pack of wild horses, a charge by a herd of steers, and an encounter with a grizzly bear were good preparation for what lay ahead.

His original intention had been to sail to Japan, but that plan was thwarted by the Russo-Japanese War, so he sailed instead to Melbourne, Australia, working his way as a deckhand. He then rode north to Sydney, to Brisbane, and eventually to Townsville in northern Queensland, with the intention of catching a boat to China. His ride through Queensland was perhaps the most daunting. First, the rains turned the black earth into glutinous mud, then he succumbed to a touch of dengue fever. In addition, his tires suffered innumerable punctures from the notorious nagoora burr, a local plant bearing three fearsome, needle-like prongs on its seeds. He walked often, carrying his bicycle overhead to cross many swollen

Figure 7.13. Karl Creelman beside the Red Bird bicycle on which he rode round the world between May 1899 and August 1901. Despite encounters with wild horses in Canada, crocodiles in Australia, and malaria in India, he never gave up on his journey. He arrived home a hero.

streams. Near Townsville, he was crossing a stream on stepping stones when a crocodile attempted to knock him off.

Yet all that effort was for nothing! Unable to find a boat bound for China, Creelman had to ride all the way back to Sydney, where he signed on a boat to Colombo in Ceylon, working as an ostler with 230 wild horses. There was time for a brief bicycle trip from Colombo to Kandy in Ceylon before the boat continued its voyage to Calcutta, where he arrived in October 1900 at the tail end of the monsoon, in time to be robbed of most of his possessions. Although the city was like a steam bath, he set off for Bombay on his trusty Red Bird. Unfortunately, as a result of riding in the cooler air of the mornings and evenings, when mosquitoes are active, he contracted malaria, and by the time he reached Benares he was in bad shape and had to spend ten weeks convalescing in the Black Watch Regiment's sanatorium.

Eventually, still in precarious health, he set off for Bombay, where he had intended to catch a boat to Cape Town, whence he would ride to Cairo, covering the length of Africa. Fortunately, no such boat could be found, and he accepted a free passage to Egypt offered by a Nova Scotian skipper. Disembarking at Suez, he partly rode but mostly walked through the desert to Cairo and Giza and then took a boat from Alexandria to Hull in England, which he reached in late April 1901. His last few weeks were spent touring the British Isles, with a side trip to Belgium, Holland, Germany, and France. Then, in August 1901 Creelman signed on as a bosun's mate on a ship leaving from Manchester bound for Saint John. He returned home to a hero's welcome in Truro, and for the next few months was much in demand as a lecturer. A few years later he moved to Winnipeg, where he married, but his health never fully recovered from his trip. At the age of only forty-six, he died in Vancouver in 1924 after a lengthy illness.

What propelled an introverted young man living in a small Canadian town to embark upon and doggedly complete such a risky project? The best clue is provided by Creelman himself. His ambition in life was to become a journalist. Lacking any obvious academic credentials, he chose to put himself in the vanguard of 1890s lifestyle experiences by cycling round the world. Moreover, he succeeded. He was welcomed back to

Truro as a pathbreaker: home-town lad conquers the world with a bicycle! Distances had indeed been compressed when the son of a working man could achieve such a goal under his own steam.

NEW VISIBLE PLACES

The new geography produced by the bicycle was characterized not only by the compression of space – to use David Harvey's term.[17] It was, in addition, strongly connected with a number of very visible places. Space is relative: the bicycle brought the world closer together; place is absolute; and bicycles were brilliantly suited to making their riders visible, provided that they were ridden in the right places. Note that, except in the learning phase, when riding academies often were visited, bicycles were infrequently ridden in private. Bicycling was usually a public exercise. Three particular places will be highlighted here; the racetrack; the photographic studio; and finally, the parks and streets where cyclists could act out the role of 'flâneur on wheels.'

The Racetrack

Bicycle races had a twofold influence on the pursuit of modernity. They promoted technological innovation, since the first prize went to the fastest machine. There were, of course, big commercial advantages to winning races. In Britain bicycle races were closely connected with invention and improvement to bicycle design. In Canada there was much less local innovation, and races served mainly to identify which imported models were the fastest and most likely to be in demand. These races also became a forum for popular culture, a type of carnival, which the movers and shakers in Canadian society attended in large numbers to enjoy one of the latest thrills.

Some of the photos used to illustrate previous chapters reflect this heightened sense of interest. For instance, despite the damage to the photographic plate, the image in figure 1.4 captures some of the excitement of seeing a highwheeler race at the Morrow Park racetrack in Peterborough. Most heads are turned, following with interest the precarious progress of the three riders, while several are standing on the railing to

Figure 7.14. A bicycle race in Portage la Prairie, Manitoba, summer 1895. The crowd by the tent in the background are waiting at the finishing line.

get a better view. In a small town like Peterborough, located on the Shield margins, this was one of those occasions on which citizens could catch up on new fashions flowing from 'the centre.'

Bicycle races continued to attract considerable interest throughout the safety bicycle era. The start of a bicycle race held in the raw frontier town of Portage la Prairie, Manitoba, in the summer of 1895 is depicted in figure 7.14; five enthusiasts line up with their trainers for a race around the local track. In figure 7.15 the start of similar race, possibly held the same summer, is shown, but this time it is a rather more auspicious occasion in Charlottetown, Prince Edward Island. The profusely garlanded hats of the ladies in the foreground and the jaunty angles of the men's headgear suggest that it is quite a classy event. A brass band (right background) was in attendance to entertain the crowd. Charlottetown's citizens evidently felt that bicycle races should be put on in style.

It is noteworthy that the bicycle boom continued for a year or so longer in Canada than in the United States, mainly because the International

Figure 7.15. The start of a bicycle race at the Charlottetown racetrack, circa 1897. Both the men and the women spectators are wearing an interesting array of hats and caps. The photo was taken by PEI's leading photographer of that period, A.W. Mitchell.

Cyclists Association's World Cycle Racing Championships took place in Montreal in August 1899. The event attracted considerable attention in the Canadian press and in social circles. It was, after all, quite a coup for a self-governing dominion to host such an event. The idea of a modern nation state had been evolving since the time of the French and American revolutions, but Canada was rather slow in throwing off the shackles of colonialism. In a small way the Cycle Racing Championships helped to build Canadian identity as an emerging modern state, especially when Canada won the team prize. Medals awarded to participants were engraved on one side with the image of a cyclist and the name and date of the event and on the other side with the emblem of the Canadian Wheelmen's Association with the Canadian beaver placed over it.[18]

The Photographer's Studio

Both the camera and the bicycle are socially constructed artefacts that evolved rapidly during the latter part of the nineteenth century. Bicycle designs reflected not only the interests of scorchers, who wanted to go faster, but also those of people who wanted a safer bicycle, a women's bicycle, a smoother riding bicycle, a tandem, a delivery bicycle and so on. Photography likewise evolved in several different ways, reflecting the diverse social purposes to which the camera was put. The studio camera became a large box, whereas smaller, portable cameras were developed for the tourist. One particular variation was the bicycle camera, the best known being the range of Poco bicycle cameras made by the Eastman Company of Rochester, New York. The camera was mounted by a clamp on the handlebars, the bicycle acting as a tripod for taking outdoor photographs of scenes that a cyclist wished to record.

The most interesting images of cyclists were generally 'cabinet' photographs taken in a photographer's studio. This is not to deny that many fascinating outdoor photos were taken, but they were usually rather spontaneous compositions, whereas studio pictures were carefully composed with a view to making a statement about contemporary culture.[19] Thus, there was an interesting synergy between the bicycle and the camera, and both triggered minor carrier waves leading to major innovation waves in the twentieth century. The Lumière brothers in the vanguard, innovators moved on from the still camera to the movie camera, video technology, and modern cinema; the links between the bicycle and the automobile have already been noted.

In order to explore the importance of the photographic studio, three images have been selected. In figure 7.16 Mr Harrison of the Ottawa Bicycle Club in August 1885 is shown with his Kangaroo dwarf high-wheel safety bicycle. Invented in 1884, this may well have been the first Kangaroo imported into Canada. Layers of modernity are visible in this picture. The photograph, which was taken in Topley's studio, was intended to suggest a sylvan setting (this tree performed sterling service for Topley in numerous pictures – for instance, it appeared four years later in a photo shown in figure 2.5). Mr Harrison is dressed in his Ottawa Bicycle Club uniform with a fashionable pith helmet similar to the one Thomas Stevens wore on his round-the-world trip begun in the previous year. His Kangaroo dwarf ordinary was introduced by Messrs Hillman, Herbert, and Cooper at the Stanley Show in 1884. This technology made

Figure 7.16. Mr Harrison of the Ottawa Bicycle Club at Topley's Studio, August 1885.

the front wheel much smaller and safer to ride – a 38-inch front wheel was geared up to 56 inches. In a race staged in England in 1884 George Smith beat the 100-mile record on a Kangaroo, lowering it to 7 hours, 11 minutes 10 seconds.[20] In figure 7.17 Mr McLeod, president of the Montreal Bicycle Club is shown on a loop-frame tricycle. A number of technological innovations were adapted for the tricycle. Not very visible is the rack-and-pinion steering controlled by rotating the right handle. There is a dome bell beside Mr McLeod's left hand and beside his right hand a vertical lever with a black handle, which activates a band brake. Tricycling was more popular among older men and among women. In figure 7.18 two young riders are presented sitting casually in front of their bicycles, on the left a Victor highwheeler and on the right an Eagle, which was a highwheel safety bicycle with the small wheel at the front to prevent headers. The very latest in technology and fashion is here being photographed, either in 1889 or 1890. The person seated on the left is R.S. (Sam) McLaughlin, whose father was proprietor of the McLaughlin Carriage Works, then the largest carriage maker in Canada. Sam went on to develop the automobile firm that subsequently manufactured Buicks under contract, before being bought out by General Motors and becoming the largest G.M. factory in Canada. His friend is seated in front of an Eagle, a highwheel safety bicycle produced only in 1889 and 1890.

Joan Schwartz, curator of the National Photographic Archives in Ottawa, has given considerable thought to the meaning of 'photographic messages' like these. Are these images simply empirical descriptions of the cycling scene, or are they creative works of art? Given the care taken in composing many studio portraits and given the commercial possibilities of a popular picture, they were clearly cultural constructs laden with meaning. For instance, there is a genre of Canadian studio photographs that were shot as a series to represent allegorical hunting or fishing adventures. Such series would involve trucking barrels of salt into the photographer's studio to recreate snow scenes. The costs would be recovered if a series became popular with the general public and sold well. The subtext of these series was usually the brave hunter conquering the hazards of nature.

Figure 7.17. Mr C.H. McLeod, president of the Montreal Bicycle Club, on his new tricycle in 1885.

Figure 7.18. Sam McLaughlin and friend with their Victor and Eagle high bicycles, circa 1889.

Bicycling scenes have a somewhat different meaning. They are geographical documents, grounded in space and time. They are also sites where 'beliefs are contested, knowledge is negotiated, and meanings are constructed.'[21] All three of the selected studio photographs of bicyclists were grounded in the second wave of the bicycle boom, which was a period of technological ferment as inventors sought to overcome the dangers of the highwheeler. Three different solutions are on display – the dwarf highwheeler (known as the Kangaroo), the tricycle, and the Eagle (with its small wheel out front). More subtly, they are grounded in space – specifically, eastern Canada. Note that each cyclist is pictured in a hardwood forest. The scene is very different from a British rural landscape with cottages and hedgerows, a prairie landscape of earth and sky, a northern landscape of pine and spruce forest, or an eastern-seaboard landscape of towns and plantations. The image is not quite 'the bicycle in the wilderness,' but more 'the bicycle in the settled frontier,' which was the reality of eastern Canada in the 1880s.

The images also contest belief in the status quo. In an era when railways, streetcars, and horse-drawn carriages were the dominant form of transport, these images throw down the cyclists' gauntlet: they offer a better way to travel. The prominent display of the machines clearly reflects the technological knowledge being negotiated – all three machines are new variations on the theme of the bicycle. Owing to gearing, the size of the Kangaroo's large wheel is much reduced; the chain drive mechanism that increases the gearing of the wheel is prominently displayed. The tricycle likewise places the chain-drive mechanism, brake drum, and steering mechanism close to the photographer. In figure 7.18, the different geometries of the two bicycles are plain to see: the ordinary and the Eagle are almost mirror images of one another. These three studio photographs speak for technical modernity and for a new and progressive bicycle age in which wheelmen stand out as technical, social, economic, and cultural leaders. They are pedalling to a brave new age.

The Flâneur on Wheels

The idea of the flâneur who took delight in the joys of the modern city as he strolled along the boulevards of Paris was central to the vision of

Charles Baudelaire in his seminal essay of 1861 on 'the Painter of Modern Life.'[22] In recent literature, the flâneur has been identified as one of the central images of cultural modernity.[23] Just as a peacock has need to display its feathers, so there was little point in purchasing the trappings of social and economic modernity if they were not strutted and displayed. During this display a great deal of observing also went on, so that the latest shifts in taste and fashion were quickly detected and acted upon. The flâneurs wandering the streets of London, the boulevards of Paris, and in New York's Central Park were, therefore, both watching and being watched. If one of the arbiters of fashion and taste were to appear in a new guise, the crowd would quickly detect this shift and follow suit.

The concept of the flâneur can be extended to include the cyclist.[24] The main phase of flânerie on the streets of Paris was said to be in decline by the 1870s, owing to three factors. First, the growth of city traffic made it increasingly difficult to wander the boulevards of large cities. Second, the city was progressively losing its mystery as a Cartesian order was placed on it – streets were widened, addresses given sequential numbers, streets lit at night, and Byzantine medieval structures replaced by large modern edifices. Third, the beginning of suburbanization was under way, with streetcars and suburban railways making family life at the edge of the city a possibility.

The classic flâneur living in Paris in the period of the Second Empire belonged to the bourgeoisie and was male. Several contributors to the volume edited by Keith Tester have argued, however, that the concept should be broadened: the flâneur might live in cities other than Paris, at times other than the 1860s, and might equally be a woman.[25] In the same vein, the possibility arises that a variant of the flâneur appeared in the 1890s, mounted on a bicycle. Several aspects of cycling at that time correspond with the original concept. Cycling was a social activity, it was often done in a crowd, it was highly visible, the rider was observed and observant, he or she had fleeting encounters with other riders and pedestrians, and the route could vary at the rider's fancy. The cycling flâneur could easily venture into the countryside, where traces of the past, which so interested the original flâneur, still could be found and the bustle of the city avoided. Like the original Parisian flâneur, the cycling flâneur was an ephemeral figure: he or she appeared in cities such as Montreal and Toronto in the early 1890s and by the late 1890s had faded from the scene. The flâneur on wheels was simply another temporary player in the drama of modernity, strongly attracted to visible places where he or she could join and interact with the crowd, places such as parks, boulevards, and the country lanes surrounding the city.

Modernity was a complex movement, advancing in sometimes conflicting ways in industry, agriculture, transport, literature and art, society, and politics. Modern industry, of itself was not always visible, but its fruits most certainly were. Wealthy consumers of the late nineteenth century took advantage of both public and private spaces to parade their new possessions. Here, the bicycle truly came into its own. Not only was it a modern artefact in its own right, not only did it come with assorted bells and whistles that marked the rider as a social innovator, not only did it command the attention of bystanders on the street with its bells and alarms, but also it could be used to parade other innovations – new styles of clothing, hats, and footwear; new parasols; new accessories; and even new cameras. Whereas the sidewalk at times could become crowded and make the flâneur invisible, a cyclist on the roadway stood out. She or he could pedal at a low speed (not much faster than a pedestrian, if necessary), but could also scorch along as fast as a galloping horse if that was the desired impression.

There were other less important places where bicycles achieved some prominence as instruments of modernity, such as the battlefield, advertisements, and factories, but the three key places identified here – the racetrack, the photographer's studio, and the streets and parks used for 'bicycle strolling' – were where the bicycle had its biggest and most enduring impact as a harbinger of modernity. Simultaneously, the bicycle occupied a key transitional position in the shift from the more circumscribed personal geographies of the mid-nineteenth century to the space-gobbling geographies of the twentieth century. Maxim makes the same argument, although his purpose is to account for the delayed appearance of the automobile: 'The reason why we did not build mechanical road vehicles before this is because the bicycle had not yet come in numbers and had not directed men's minds to the possibilities of independent long-distance travel over the ordinary highway. We thought the railroad was good enough.'[26] The ability of the bicycle to cover distances quite rapidly had the net effect of initiating a new and modern geography. Certainly, as Hounshell stresses, bicycle production paved the way for automobile production, but it also fostered a new geography of the mind. It created 'the desire for swift and cheap personal transportation.'[27] In the twentieth century, not only has this shrinkage of space continued, but so also has the ability of travellers to set their own itineraries.

CHAPTER EIGHT

Pedaller's Progress: The Bicycle and Modernity

In this final chapter the analysis is reversed, to reflect on the insights that the preceding exploration of Canada's bicycle era may add to our understanding of modernity. To the extent that modernity has penetrated almost all aspects of everyday life, a study of its practical impacts ought to provide a useful basis for such an assessment. The average person has not experienced modernity as a cerebral process, and even fewer see it as a meditation upon the philosophy of the Enlightenment project. Indeed, it seems highly likely that many who embrace modernity have never read its foundational philosophers. The majority are engaged in the project as an exercise in the development and application of rational and superior principles – learned mainly by seeing and experiencing them – that are designed to displace existing and supposedly inferior practices in agriculture, industry, medicine, government, and numerous other aspects of contemporary life.

Commentaries on modernity often present material progress as a universal process. Nigel Thrift, in contrast, contends that the presumption of universality is one of the great myths of the Enlightenment project.[1] Modernity, he suggests, is culturally (and therefore geographically) specific. It follows that modernity is not a universal process, but a complex one woven around things locally and regionally embedded. In this study the emphasis has been on practical matters of innovation, design, mass production, mass consumption, improved infrastructure, and social practices in specific Canadian settings. Although the bicycle era was only a brief episode in the complex drama of modernity, it provides a number of insights into the workings of the broader cultural movement. The importance of the philosophical roots of modernity is not in question. Most movements are founded by philosophers who have that rare ability to discern broad patterns of development well in advance of the mainstream. Bentham and Marx perceived the social and economic consequences of industrialization long before socialist parties were formed. Thoreau and George Perkins Marsh worried about environmental destruction ages before the Sierra Club and Green parties were launched. Indeed, many of the world's major religious and cultural movements have obscure philosophical origins. They have subsequently become what Jean Baudrillard calls 'modes of civilization' by developing a material form with a popular local following.[2] In so doing, small groupings with abstract philosophies have been transformed into diverse mass movements.[3] In the case of modernity, a series of decisive historical events, most notably the American and French revolutions, coupled with rapid advances in agriculture and industry, gave concrete and local form to the abstract notions of the Enlightenment, in the process engaging large sections of western civilization in advancing modern life. Indeed, modernity in its many diverse forms has so captivated the public's imagination over such a long period that it has had a decisive influence on recent civilization. During its turbulent (and still incomplete) history, it has passed through many phases as new technologies and fashions were discovered, modified, diffused, and then discarded, one after another, each in turn being replaced by new ideas that revived the project, and relaunched it in new directions.

One such phase was the bicycle era, which, even though it was a relatively minor sideshow in the overall drama, nevertheless illuminates several important aspects of modernity that have not been stressed in more abstract discussions. This conclusion will therefore draw on the preceding analysis of the bicycle age in Canada to highlight four aspects of modernity. First, modernity is antagonistic to the status quo. Second, and paradoxically, although local traditions form a part of the established practices that are inimical to modernity, modernity itself becomes locally embedded. Thus, the process of creation and destruction that accompanies the march of modernity applies not only to the technologies and ideas being diffused, but also to the way they are locally adopted. Third, a part of this geographical embeddedness has resulted from the ability of modernity to seduce the 'crowd,' in slightly different ways in different places, mainly because it is able to create spectacles that have a local resonance. Finally, the practical manifestation of modernity has been most firmly based in industry; the abstract rhetoric of rationality and reason has found its most decisive incarnation in factories, in the industrial workforce, and in the production of a succession of consumer goods of varying utility. The essential subtext to the discussion is that modernity has become massively important precisely because it grew from its metaphysical origins into a popular and diversified movement that progressively infiltrated every practical aspect of western civilization.[4] On the economic side, this infiltration took place through a succession of carrier waves; in fact, such waves are seen to provide the key to understanding the dynamics of the entire project.

CHALLENGING THE STATUS QUO

Modernity stands in opposition to established traditions and practices. Traditional forms, which of their very nature are locally embedded, try to resist modernity. Absolute monarchs, the papacy, guilds, Shintoism, masonic lodges, communal and hereditary landholders, male chauvinists, and many other hierarchical groups thrive on tradition and actively resist change. Modernity undermines the authority of tradition because it empowers a new elite that has access to technology, finance, and rational knowledge. By continually fostering demands for novelty, 'modernity gives rise to an aesthetic of rupture, of individual creativity, of innovation marked by the sociological phenomena of the avant-garde and by the always more extensive destruction of traditional forms ... [it] accelerates the destruction of the indigenous way of life and precipitates social demands for change.'[5] Modern cultural forms do not always try to overthrow traditional forms directly, but they will often absorb established local practices to avoid conflict.

Cyclists, even though they developed their own traditions, such as cavalry practices in the age of the highwheeler, sought to change other aspects of late-nineteenth-century society. They challenged the supremacy of traditional users of roads and parks, and when they met with resistance from teamsters and the horse lobby, they took up the challenge in newspapers and journals, in council chambers and provincial legislatures, and in the courts. They advocated the use of the bicycle as a way of modernizing warfare and met with considerable resistance from the conservative military establishment. Women cyclists were prominent among groups rejecting strict Victorian dress and exercise codes. Male cyclists used bicycle clubs to resist the authority of the older generation. When it came to shaking up traditional systems of constructing and maintaining roads, cyclists again were among the more vocal agitators for change. Overturning artisanal traditions, a few bicycle manufacturers broke through to mass production to create a large industrial workforce. The selling of bicycles pushed the commercial bounds of modernity onto new ground, using some extraordinarily effective imagery in advertising and sponsoring races and long-distance journeys. The evidence of the bicycle movement confirms that a central characteristic of modernity was its opposition to the status quo.

The bicycle era also illustrates many of the complexities of modernity, its tendency to destructive excess, its ability to fragment into rival subgroups, and the development of its own competing local traditions. By the end of the bicycle era, some of the more destructive aspects of this rupture with tradition were beginning to appear. The dehumanizing effects of the mass production line, the concentration of industrial ownership, the introduction of new models for their own sake without any significant technological innovations, and the growing emphasis on mass marketing all point to aspects of modernity that are largely bereft of the original inspiration of reason and progress.

LOCAL EMBEDDEDNESS

The emphasis placed here on the way modernity breaches tradition connects with an important criticism made by Anderson of Marshall Berman's *All That Is Solid Melts into Air*.[6] Anderson contends that Berman and a number of other writers come close to espousing technological determinism when they equate modernity with technological change. Phil Cooke focuses particularly on this criticism and the way Berman tends to leave out the people who produce and consume modernity.[7] The analysis of Canada's bicycle boom presented here not only highlights the people who invented, produced, and consumed cycles, but also speaks to the privileged class that led the bicycle carrier wave, argued for the politics of the bicycle, and sought to stamp out resistance to bicycling from traditionalists. In the final analysis, the only recognizable tradition of modernity was 'the tradition of the new.'[8]

There follows a very important geographical corollary to this point. For if modernity was simply a matter of technological and social diffusion, then it would have a homogenizing effect. Berman stresses modernity's universalizing aura, and Baudrillard is unequivocal: 'confronting the geographic and symbolic diversity of [tradition], modernity imposes itself throughout the world as a homogeneous unity.'[9] Paul Rabinow, in his study of modernity in France, makes a similar point in suggesting that 'traditionalist and reactionary currents valorized differences, pluralism, and localism, in opposition to what they understood as the centralizing, levelling influences of the Jacobin state.'[10] But Rabinow proceeds to qualify his position in a very important way. He notes how the French defeat at the hands of the Prussian armies in 1870, which was attributed in part to a lack of training in map reading and a poor understanding of local terrain by French officers, caused a profound shock in French geographical science. The Prussian success was attributed to the superiority of their science, which could be answered only by upgrading French science and technology, including the geographical sciences. The point is that, although the technological dimension of modernity was, in principle, universal, it was guided and regulated both by individuals and by the state, which imposed local understandings and practices on the process.

This local contingency was borne out by the bicycle, which took on subtly different uses and meanings in all the regions affected by the cycling wave. The vast majority of Canadians cyclists were active only during the season from May to October. Unlike their British and French counterparts, until the 1890s they took very little interest in developing and manufacturing these bicycles. Very few French-Canadians, the majority of whom then formed a traditional rural society, took to riding these early cycles; it was Anglo-Canadians who initially dominated Canadian cycling. Canadian invention and innovation related to the bicycle was heavily concentrated in the major towns and cities of southern Ontario. It was also in Ontario that several firms mass produced safety bicycles, but with distinctive Canadian modifications, including the widespread use of wooden rims and a fascination with ice velocipedes. The point does not need to be laboured; for it is clear that, although technology itself has a universalizing effect, invention, investment decisions, the human administrative and decision-making systems that manage technology, and the culture of production are locally specific. The net effect is that local elements negotiate backwards and forwards with the universal elements of modernity. There is a further corollary to this point. Berman's interpretation of modernity stresses technological change and the endless quest for things new, but he overlooks the fact that this process extends to our grasp of modernity itself, which is constantly mutating to produce competing visions and understandings. There were, for example, two quite different visions of modernity guiding the factions in the Toronto Sunday streetcar debate.

MODERNITY AS A SPECTACLE

Many of the activities of nineteenth-century cyclists were intended to show cycling as a spectacle. When they were riding their highwheelers in club uniforms in formation down the main street, racing in front of a throng at the racetrack, gathering in groups in public parks, or travelling abroad, cyclists wanted to be seen, and were very conscious of how they appeared to the larger public. Like media stars a century later, they felt it was important to be seen frequently, to be seen to progress either in technology or fashion, and to update one's image regularly. In consciously creating specific images, which were often recorded on film, bicyclists were writing a particular history of modernity. Conspicuous in the public spectacles they mounted were new bicycles, new accessories, new clothing, new places visited, the companionship of cycling, and the dignity bestowed by a bicycle on its owner.

Whereas today antique bicycle enthusiasts look backwards in time and justify their interest in historical terms, the cyclists of the nineteenth century looked mainly to the future. It is interesting to note that they also had established traditions such as club etiquette, hierarchy, seasonal activities, and dress codes, which point to the existence of subtle countercurrents within the broad tide of modernity. In the case of bicycles, the appeal of technological modernity lay in the very visible progress made in a quarter-century and in the promise of further advances. The succession of late-nineteenth-century bicycles, which became progressively safer, lighter, more comfortable to ride, more adaptable, and faster, gave legitimacy to the costly quest of research and innovation. Habermas has argued that modernity always lived for the future and for the promise of novelty; the spectacles associated with bicycling provide evidence that this indeed was the case.[11]

INDUSTRY AND MODERNITY

Industrial capitalism was required to transform the intangible concepts of Enlightenment philosophy into a popular movement in pursuit of modernity. Krishan Kumar is quite emphatic on this point: 'modernity and industrialism are closely if not intrinsically related. Our very image of modernity is formed to a good extent of industrial elements. It is difficult to think of the modern world without conjuring up steel, steam and speed. From the Great Exhibition of 1851 in England to the World's fairs of the 1930s in America, industrialism trumpeted its achievements and pronounced itself the salvation of humanity. The great cities of modernity, especially American cities such as New York and Chicago, are inconceivable without industrial technology.'[12] Industrial modernity in Canada was not advanced by the bicycle until quite late, and even then its impact was restricted very largely to southern Ontario. Despite a growing interest in cycling in Canadian towns in the 1880s, production was confined to a few artisanal workshops. This situation changed in the 1890s, however, with the appearance of the more popular safety bicycle, and Canadian firms began to manufacture a range of good-quality machines. By the late 1890s the adoption of mass production methods had lowered the unit cost of a bicycle to a point where a much wider spectrum of Canadian society could afford to purchase one.

The engraving of the Massey-Harris bicycle factory shown in figure 3.7 illustrates the importance attached to size within the dominant contemporary discourse on industrial modernity. In the industrial age, scale and speed became defining indices of progress. The longest bridges, fastest ships, deepest mines, and highest buildings were heralded as 'wonders of the modern world.' Engineers in the tradition of Brunel, de Lesseps, and Edison sought continually to push the limits of technology. For many, large factories and big corporations were viewed as progressive forces that marked the passing of more primitive forms of artisanal production. Artists, therefore, often magnified the dimensions of late-nineteenth-century factories, such as the Massey-Harris bicycle factory, in order to exaggerate the transition to mass production and promote a sense of awe about the scale of its operations. Yet there were counter-currents to this version of modernity, particularly within the competing frameworks of political modernity and democracy in the workplace. These alternative voices expressed scepticism about the advantages of largeness and commercial domination, because they threatened modern understandings of individual rights and social justice. Big corporations, monopolies, and cartels had the power to limit consumer choice, set prices to suppliers and customers, and restrict workers' rights. Subsequent anti-trust legislation enshrining modern democratic principles in the market place and curbing the powers of the firm, grew out of understandings of modernity quite different from those of the Carnegies, Rockefellers, and Popes of the corporate world, who viewed size as synonymous with progress.

THE END OF THE RIDE

In Canada, interest in bicycling declined after 1898, less because of a decline in sales – indeed, sales recovered in the new century – but because consumers on the cutting edge of fashion became fascinated with a new set of innovations. The bicycle carrier wave had run its course. The herd instincts of the crowd led it to follow the next wave of innovators. In effect, the bicycle had become a standardized, rather boring, mass-produced item, which no longer bestowed much kudos on its owner. There was nothing new about this process. Modernity has always had its most sustained impact on the world through technological innovation and the resulting changes wrought in manufacturing, transport, military methods, and the provision of services. From the time of the Industrial Revolution through to the present, modernity has unfolded through a relentless sequence of innovation and rejection.[13] This process has been nourished by a belief in the technological marvels of industrial civilization. The trajectory of innovation has not been a simple linear one; on the contrary, it has moved in diverse ways in different contexts, and the various strands of modernity quite often have been in opposition to one another. Modernity as a whole, however, has prevailed over counter-movements during two centuries of remarkable scientific and economic progress. Indeed, the single most important understanding of modernity to be drawn from this examination of Canada's bicycle boom is that, in its material form, it has found expression as a series of carrier waves. Such a conclusion draws the subject even more closely into the geographer's sphere of interests, since a carrier wave diffuses through time and over space, from the centre to the margins. In consequence, modernity is also closely linked with geographical issues of territorial development.

The argument can best be illustrated by borrowing the wonderful metaphor of *cultural seismology* developed by Bradbury and McFarlane.[14] They distinguish displacements of three orders of magnitude. First, there are minor tremors associated with innovations that trigger a brief ripple of interest but then soon fade away. Second – and it is at this level that the bicycle belongs – there are innovative eruptions whose effects last much longer and that ripple into neighbouring areas. The impacts may be felt for several decades and over a large area. Finally, there are clusters of innovations that constitute major fault lines in economic development, innovations exclusively identified with the propulsive industries that drive long waves of economic development. Although the carrier wave associated with the bicycle was felt at the second order of magnitude, it ranks even higher on the economic Richter scale, because the innovations it set in motion carried through to the automobile and the major longwave initiated by it.

Techno-modernity advances by way of a succession of carrier waves in the manner described at length above. First, there must be an act of invention that catches the attention of, and is adopted by, a group of social leaders. This initial success triggers a cluster of inventions that improve upon the prototype and spread into a number of related activities. Rapid development will occur not only in the primary innovation, but also in the development of related accessories. Certain versions of the innovation attract sufficient interest that manufacturing begins, and as the scale of production increases, unit costs decline. The manufacture of a series of related products and accessories follows soon after. Consumption of the innovation is stimulated by marketing and advertising, so that demand for the product grows, allowing manufacturers to benefit from increasing returns to scale. Widespread acceptance of the product may result in bottlenecks in related infrastructure (roads for cars and bicycles; satellites for radio communications). Such carrier waves do not achieve universal acceptance but are worked out in geographically diverse ways. The product of a carrier wave is normally adopted for use in particular spaces and for display in specific places. Thus, each carrier wave is likely to produce a new geography of modernization. Eventually, all carrier waves run out of steam, to be replaced by a new wave of innovations.

So it was with the bicycle. The total number of bicycles in Canada actually rose after 1900, but this increase was no longer important to the project of modernity. It was rather like the decision made by Statistics Canada that in the 1991 Census refrigerators and televisions would no longer be counted; since most households by then had these appliances, they had ceased to be a useful social marker. By 1900 the bicycle no longer distinguished its owner as a pioneer of things modern. The boom was over.

Notes

One: Modernity and the Bicycle

1 This anecdote is taken from Heather Watts, *Silent Steeds: Cycling in Nova Scotia to 1900*. Watts, formerly of the Nova Scotia Museum in Halifax, dates this event to 1866, but since the velocipede was hardly known in England before 1868, 1869 seems a more probable date. The source was an article published in the *Cape Breton Mirror* 2, 11 (October 1953), in which the date, 1865, is suggested.

2 Details of how this photo was taken can be found in Constance Kerr Sissons, *John Kerr*.

3 Kerr later worked at Notman's photographic studio in Montreal, went on to hunt buffalo on the prairies and served with the forces of the Red River Settlement, where he opposed the Métis rebellion under Louis Riel, before returning to Perth to serve as town clerk for many years. An account of the long and adventurous life of John Kerr can be found in Sissons, *John Kerr*.

4 Thanks to the contributions of the late Winger Thomas, the part played by bicycles in the history of Stratford is well documented in Adelaide Leitch, *Floodtides of Fortune: The Story of Stratford and the Progress of the City through Two Centuries*.

5 Keith Walden, *Becoming Modern in Toronto*, 4.

6 Several excellent technical histories of the bicycle are already available, including G. Donald Adams, *Collecting and Restoring Antique Bicycles*; Nick Clayton, *Early Bicycles*; Pryor Dodge, *The Bicycle*; Andrew Ritchie, *King of the Road*; and Derek Roberts, *The Invention of Bicycles and Motorcycles* and *The Invention of the Safety Bicycle*. The Canadian scene is described by Sharon Babaian, *The Most Benevolent Machine: A Historical Assessment of Cycles in Canada*. Although little known, Leslie Dunham's Harvard PhD thesis, 'The Bicycle Era in American History,' is an excellent research

document on early American cycling history.

7 Eugen Weber, *My France: Politics, Culture, Myth*, 11. These ideas were developed earlier in Eugen Weber, *Peasants into Frenchmen: The Modernization of Rural France 1870—1914*. A full and fascinating account of cycling in modern France is given by Christopher S. Thompson in his PhD dissertation, 'The Third Republic on Wheels: A Social, Cultural and Political History of Bicycling in France from the Nineteenth Century to World War II.'

8 Details are given in Glen Norcliffe, 'Popeism and Fordism: Examining the Roots of Mass Production,' and 'Colonel Albert Pope: His Contribution to Bicycle Manufacture and the Development of Mass Production.'

9 Clark, *The Painting of Modern Life*, is seen as one of the seminal essays on modernity.

10 Subsequently, and appropriately for a leading interpreter of modernity, Zola became one of the 'fervants de la pédale.' He took afternoon rides 'pour la joie de l'exercise et du plein air'; *Info-Vélocithèque*, 30 (November 1997), 1–2.

11 Note that I do not include the hobby horse (or draisine) as a true bicycle: I prefer to classify it as a walking machine, and use the term 'bicycle' for a two–wheeled machine propelled by a rider who does not touch the ground.

12 During the 1990s David Herlihy has investigated the early history of the bicycle. See, especially, Herlihy, 'Lallement vs. Michaux: Who Truly Invented the Bicycle?,' and 'H. Cadot: one of the earliest bicycle makers in France.' The roles of Michaux and Lallement and of the Olivier Brothers nevertheless remain a bone of contention. See, for instance, Roland Sauvaget 'Michaux/Lallement'; his update of that article, 'Michaux-v-Lallement: The Conclusion?'; and Hans-Erhard Lessing, 'Around Michaux: Myths and Realities.'

13 G. Lacy Hillier, 'Cycles Past and Present.'

14 These claims were made by Lallement in the testimony he gave under oath in the case of Pope vs McKee and Harrington. See Herlihy, 'Lallement vs. Michaux,' for details.

15 Andrew Ritchie, *King of the Road*, 54–7.

16 See Hans-Erhard Lessing (n.d.) 'Karl von Drais' Two-Wheeler – What We Know'; and *Karl von Drais: der Empire-Technologe Wird Rehabiliert*; and Roger Street, *The Pedestrian Hobby-Horse: At the Dawn of Cycling*.

17 Miles Ogborn, *Spaces of Modernity*, 1–2.

18 Philip Cooke, *Back to the Future*, 7.

19 The definition of modernity in Derek Gregory, *Dictionary of Human Geography*, 388.

20 Perhaps the most famous of these scientific societies is described in Robert E. Schofield, *The Lunar Society of Birmingham*.

21 Miles Ogborn, *Spaces of Modernity*. Ogborn identifies Zygmunt Bauman, *Modernity and Ambivalence* as a writer who stresses the search for rationality, in contrast to Marshall Berman, *All That Is Solid Melts into Air*, (who highlights the endless fluxes of modernity. Pred and Watts's notion of 'symbolic discontent' captures this sense of flux; see Allan Pred and Michael J. Watts, *Reworking Modernity: Capitalism and Symbolic Discontent*.

22 See Paul Glennie and Nigel Thrift, 'Modernity, Urbanism, and Modern Consumption';

and 'Consumers, Identities, and Consumption Spaces in Early-Modern England.'

23 In the view of Bryan Turner of Cambridge University, there were three main strands of modernity: the political, the technological cum economic, and the social/cultural. Public lecture at York University, Toronto, 14 April 1999.

24 Gregory Baum, 'Modernity: A Sociological Perspective,' 3.

25 These details were gleaned from an 1819 article entitled 'The Pedestrian's Accelerator.'

26 Street, *The Pedestrian Hobby-Horse*, presents the most comprehensive study of the hobby horse. See also Derek Roberts, *Cycling History: Myths and Queries*, 10–11.

27 Roberts, *Cycling History*, errata sheet for p.10.

28 See Françoise and Serge Laget, *Le Cyclisme*, 10.

29 Contemporary reports in Canadian broadsheets give the impression that their authors had seen the machine first hand, but I have yet to find an article in which the presence of a hobby horse in Canada is specifically confirmed. My assumption is that they were seen briefly in Halifax, Quebec City, Montreal, and possibly Kingston.

30 Pryor Dodge, *The Bicycle*, 20 suggests that the hobby horse was popular only from the summer of 1817 to late 1819.

31 I examined microfiche of the following broadsheets: the *Upper Canada Gazette*, the *Niagara Spectator*, the *Kingston Chronicle*, the *Halifax Weekly Chronicle*, the *Acadian Recorder*, and the *Halifax Free Press*.

32 A Canadian broadsheet in 1819 reported: 'the velocipede ... promises to be as much the rage as the kaleidoscope was of late,' *Kingston Gazette* 1, 31 (30 July 1819), 2, c. 4.

33 Weibe E. Bijker, *Of Bicycles, Bakelites and Bulbs: Towards a Theory of Socio-Technical Change*; and Trevor J. Pinch and Weibe E. Bijker 'The Social Construction of Facts and Artefacts: Or How the Sociology of Science and the Sociology of Technology Might Benefit Each Other.'

34 Street, *The Pedestrian Hobby-Horse.*

35 A six-verse poem that appeared in the *Acadian Recorder* on 2 October 1819, 4, c.1, under the title 'The Velocipede Race,' gives some sense of the fun poked at the hobby horse. Its first and last verses read: 'Rare sport for today/Come, hasten away,/ To the field of great Mars let us toddle/ A fig for the chace/ the Velocipede race/ To see all the world's on the waddle!// No velocipede came,/But who was to blame,/ The wheels were in want of new spokes;/ And by setting of sun,/The devil a one,/ But knew, he'd been had by the hoax.'

36 See Edward Herman and Noam Chomsky, *Manufacturing Consent.*

37 Gregory Baum, 'Modernity: A Sociological Perspective,' 3–9.

38 Clark, *The Painting of Modern Life*, 9.

39 A major influence in the United States, and in some Canadian clubs, was the 1880 book by Charles E. Pratt, *The American Bicycler: A Manual*, which lists bugle calls, club rules, and courtesies of the road; it draws on his own experiences in the American Civil War. Parades by these early clubs had a carnival-like quality to them; the relationship of the carnivalesque to modernity is explored in Rob Shields, *Places on the Margin: Alternative Geographies of Modernity.*

40 Max Weber, *The Protestant Ethic and the Spirit of Capitalism.*

41 Berman, *All That Is Solid Melts into Air*, chap. VI, 'Robert Moses: The Expressway

World.'

42 J. Nicholas Entrikin, *The Betweenness of Place: Towards a Geography of Modernity*. See also R.C. Harris, 'Power, Modernity, and Historical Geography.'

43 Thimonnier's original patent was granted by the French government in 1830.

44 James Naremore and Patrick Brantlinger, eds., *Modernity and Mass Culture*; and Robert David Sack, *Place, Modernity, and the Consumer's World.*

45 Henri Lefebvre, *The Production of Space*. The term 'gendered spaces' implies places some of which are largely associated with women and others with men.

46 Discussions of the product cycle may be found in R.A. Erickson, 'The Filtering Down Process: Industrial Location in a Non-metropolitan Area'; and J.M. Utterback, 'The Dynamics of product and process innovation in industry.'

47 Jonathan Gershuny, and Ian Miles, *The New Service Economy.*

48 Norcliffe, 'Popeism and Fordism,' Table 1.

49 Nicholas Oddy, 'The Bicycle – An Exercise in Gendered Design.'

50 Phillip Mackintosh, '"Wheel within a Wheel": Frances Willard and the Feminization of the Bicycle.'

51 Streetcar suburbs are discussed by David Ward, 'A Comparative Historical Geography of Streetcar Suburbs in Boston, Massachusetts and Leeds, England: 1850–1920'; and Sam Bass Warner Jr, *Streetcar Suburbs: The Process of Urban Growth in Boston 1870–1900.*

52 Both James J. Flink, *The Automobile Age*, and H.P. Maxim, *Horseless Carriage Days*, discuss the rise of the automobile.

Two: The Bicycle Carrier Wave

1 These very words, for example, are being written on an 'old' word processor that my son tells me is a technological embarrassment.

2 Marshall Berman, *All That Is Solid Melts into Air*. In practice, there may be a need for a third face, since many innovations have only cosmetic effects. See also Antoine Compagnon, *The Five Paradoxes of Modernity*; and Derek Gregory, 'Interventions in the Historical Geography of Modernity: Social Theory, Spatiality and the Politics of Representation.'

3 Dunlop was not the originator of the pneumatic tire. It had previously been patented in 1848 by Thompson, but at that time the quality of rubber was not sufficiently consistent to make a reliable tire, nor were its commercial applications as evident as they were in the 1880s. See Archibald Sharp, *Bicycles and Tricycles: An Elementary Treatise on their Design and Construction*, 159–60.

4 It is likely that many of the encomiastic letters endorsing products were forged, or at least arranged.

5 These machines were widely advertised in the cycling magazines of the 1880s and 1890s. A good selection of these bicycles is illustrated in Frederick Alderson, *Bicycling: A History*; H.W. Bartleet, *Bartleet's Bicycle Book*; Andrew Ritchie, *King of the Road: An Illustrated History of Cycling*; and Pryor Dodge, (1996) *The Bicycle.*

6 For instance, some of the times claimed for various distances on the Star Bicycle in

the catalogue of 1886 seem extraordinarily fast and have not been approached in recent times, even on modern paved roads. The claim for one mile in 2 minutes 41.4 seconds implies an average speed of 22.4 mph. As an occasional Star rider, I find this claim difficult to believe.

7 These long waves, also known as Kondratieff waves, were described by the Soviet economist, Nikolai Kondratieff, during the 1920s. His ideas fell out of favour during the boom following the Second World War, but interest in long waves revived during the recessionary years of the 1970s and 1980s. See Mick Dunford and Diane Perrons, *The Arena of Capital*; and Peter Hall and Paschal Preston, *The Carrier Wave: New Information Technology and the Geography of Innovation, 1846–2003.*

8 Dunford and Perrons, *The Arena of Capital*; Hall and Preston, *The Carrier Wave.*

9 Alexander Graham Bell's telephone, which, like the bicycle, bridged the second and third Kondratieffs also had a crucial technological precursor: von Hemholtz's discovery that electricity could be used to make forks vibrate and create sound. The telegraph, meanwhile, had prepared society for the possibility of long-distance spoken communication.

10 In Karl Kron, *Ten Thousand Miles on a Bicycle*, 790, it is reported that there was a club in Winnipeg by 1887.

11 G.C. Allen, *The Industrial Development of Birmingham and the Black Country, 1860–1927*; Andrew Millward, 'The Genesis of the British Cycle Industry, 1867–1872' and 'The Cycle Trade in Birmingham 1890–1920'; André Vant, 'L'industrie du cycle dans la région stéphanoise'; and *L'industrie du cycle dans la région stéphanoise*; David Hounshell, *From the American System to Mass Production, 1800–1932: The Development of Manufacturing Technology in the United States.*

12 The case for Lallement is made by David V. Herlihy, 'Lallement vs. Michaux: Who Truly Invented the Bicycle?,' and 'H. Cadot: One of the Earliest Bicycle Makers in France.' The case for Michaux is made by Keizo Kobayashi, 'Les Frères Olivier'; and by Roland Sauvaget, 'Michaux-v-Lallement: The Conclusion?'

13 Bruce Epperson, 'Failed Colossus: Albert A. Pope and the Pope Manufacturing Company, 1876–1900.'

14 Glen Norcliffe, 'Popeism and Fordism: Examining the Roots of Mass Production.'

15 Karl Polanyi, 'The Economy as Instituted Process.'

16 Mark Granovetter, 'Economic Action and Social Structure: The Problem of Embeddedness.' See also Glen Norcliffe, 'Embedded Innovation: Canadian Cycle-Related Patents 1868—1900.'

17 One of Canada's leading bicycle enthusiasts, Lorne Shields, has a high bicycle with the nameplate of Thomas Fane and Company of Toronto attached to it, but it is not clear whether it was made by another manufacturer and sold by Fane, or assembled by Fane using imported parts, or even fabricated by Fane. The same issue arises in the case of a highwheeler bearing the name badge of the Goold Bicycle Company of Brantford.

18 The term 'penny-farthing,' although today the best known of these names, is actually a neologism, probably dating to 1891; see Derek Roberts, Cycling History: Myths and Queries, 1. Many bicycle historians disapprove of the use of this term, even though it has higher communicative value than the other terms.

19 See Nick Clayton, 'Who Invented the Penny-Farthing?'
20 Nick Clayton, 'The Development of the Suspension Wheel.'
21 Roberts, *Cycling History*, 2.
22 Other riders dispute this claim. My own experience, since I have ridden long distances on both types of wheel, is that the difference is quite minor, although I believe that severe jolts are slightly damped by tangential spoking.
23 Sharp, *Bicycles and Tricycles,* 160.
24 The giraffe bicycle is discussed by Sharp, *Bicycles and Tricycles*, 159.
25 Sharp, *Bicycles and Tricycles*, 313–15. To alleviate these forces, later cross-frame models added stays from the crank bracket to the rear wheel spindle and from the steering head to both the seat stay and the bottom bracket, as can be seen in figure 2.5.
26 In the late 1860s a few velocipedes were fitted with rubber tires.
27 Sharp, *Bicycles and Tricycles*, 369.
28 Dodge, *The Bicycle*, 60.
29 The Pope Manufacturing Company developed the Columbia adjustable bearing, which was sufficiently different from the Aeolus bearing that the firm obtained a patent for it in 1880.
30 Better metallurgy was important to this development. In particular, tool steel, which has about 1 per cent carbon, can withstand much greater stress in a bearing than mild steel.
31 See Hillier, 'Cycles Past and Present,' who suggests that Messrs Mehew of Chelsea may have pioneered the direct drive of a front wheel by cranks. See also Keizo Kobayashi, 'The Velocipede in France, from Karl von Drais to Pierre Michaux, 1818–1861' and *Histoire du vélocipède de Drais à Michaux: Mythes et réalités*. Alderson, *Bicycling*, 22 states that both Karl Kech of Munich and Philip Fischer of Schweinfurt attached pedals and cranks to bicycles well before Michaux and Lallement did, but in neither case did the innovation diffuse or have any commercial impact. Indeed, there is a report of a Yarmouth (England) mechanic planning to attach cranks to a boneshaker in 1819; see Roger Street, *The Pedestrian Hobby-Horse*, 143–4. See also Jost Pietsch, 'The Karl Kech Bicycle – A Reassessment.'
32 From a modernist perspective, issues of historical priority in invention are of little importance unless the first prototype of a commodity captures the public imagination and enters into commercial production, that is, achieves some cultural or economic significance. By the definition adopted here, the draisine or hobby horse was a walking machine, not a bicycle. Flimsy evidence suggests that the first rider-propelled bicycle might have been Kirkpatrick Macmillan's machine, possibly copied a few years later by Gavin Dalzell; see Alastair Dodds, 'Kirkpatrick Macmillan, Inventor of the Bicycle: Fact or Heresay?' However, Gompertz's machine may also have a valid claim; see Les Bowerman, 'Lewis Gompertz and His Addition to the Velocipede.' Indeed, so might Barthélémy Thimonnier; see M.B. Gougaud, 'Note sur Barthélémy Thimonnier,' 143. None of these machines was produced in any number, and at the time they had no impact outside their immediate locality. I concur with Oddy's position when he states: 'Macmillan is, in overall terms, unimportant to a cultural historian dealing with cycling history ... in historical terms, undue weight

is given to mechanical and structural oddities ... if cycling history is to develop into a serious academic area of study, then more of its exponents must ... start looking at much broader historical issues'; see Nicholas Oddy 'Kirkpatrick Macmillan, the Inventor of the Pedal Cycle or the Invention of Cycle History,' 27–8.

33 The Lallement patent was sold to Calvin Witty of New York City in 1869. See Donald Adams, *Collecting and Restoring Antique Bicycles*, 23.

34 Other variations on the shaft-drive mechanism were marketed, including Ivor Johnson's roller pin gear and the Victor 'spinroller' of the Overman Company.

35 I saw disk brakes on an early bicycle sold at a bicycle jumble sale in Mystic, Connecticut, in 1992. It was purchased by Loren Hofstetler for his museum in Edinburgh, Scotland.

36 William S. Kelley (misspelled in the register as Kelly) was vice-president of the H.B. Smith Machine Company; see William C. Bolger, *Smithville: The Result of Enterprise.*

37 Jim Fitzpatrick's book, *The Bicycle and the Bush*, provides a revealing account of the importance of bicycles in Australia and echoes the argument made here that they were briefly of greater significance than is generally credited.

38 Fitzpatrick, *The Bicycle and the Bush*, 36.

39 Chapter four of Fitzpatrick's book, entitled 'Environment and adaptation,' is particularly good on the adaptation of the bicycle to conditions and uses in the Australian outback.

40 The extensive linkages of one of the agricultural machinery firms in Brantford is described by Gordon Winder, 'The North American Manufacturing Belt in 1880: A Cluster of Regional Industrial Systems or One Large District?'

41 This tandem is similar to Rucker and Winterschladen's U.S. patent (#3628), which Adams, *Collecting and Restoring Antique Bicycles*, 118, dates to 1883, so this may be a pirated idea rather than an independent invention, in which case it shows the importance of imitation.

42 The same is true of Massey-Harris, Canada's largest agricultural machinery manufacturer, which also became a major bicycle fabricator in the 1890s.

43 See J.J. Brown, *Ideas in Exile*; and Carole Precious, *Thomas Carbide Willson.*

44 These photographs were discovered in an attic and sold at auction in 1999. A set of ten studio photographs formed part of a promotional folder. Presumably, this is a prototype of the umbrella.

45 Marmaduke Matthews, a landscape painter, was an interesting figure in Ontario society at that time. He proposed the development of Wychwood Park, one of Toronto's most exclusive housing enclaves, and he fostered Canadian art through the Royal Canadian Academy and the Ontario Society of Artists, in addition to his bicycle experiments. See Elizabeth Wilton, 'Cloud Bound: The Western Landscapes of Marmaduke Matthews.'

46 In 1871 the population of Montreal was 107,255, compared with 56,092 for Toronto. By 1891 these figures were 216,650 for Montreal and 181,220 for Toronto.

47 These wooden-frame bicycles were very flexible and whippy to ride – indeed this was their main drawback.

48 'Velox,' 111–12.

49 See Terrence Cole, *Wheels on Ice.*

Three: Modern Manufacturing

1 Krishan Kumar, *From Post-Industrial to Post-Modern Society*, 82.
2 A.E. Harrison, 'The Competitiveness of the British Cycle Industry, 1890–1914.'
3 See William Humber, *Freewheeling: The Story of Bicycling in Canada.* Humber states (p. 50) that highwheelers were manufactured in Canada by Semmens, Ghent and Company of Burlington, Ontario (producing nickel-plated bicycles in 1882) and by Thomas Fane's Comet Bicycle Company of Toronto (probably by 1888). A few ordinary bicycles with a Fane nameplate have survived, but I have not seen any evidence corroborating the case of Semmens, Ghent which probably falls into the category of artisans. Some distinctive features of the Fane highwheeler suggest that parts of the bicycle may have been genuinely Canadian. The Goold Bicycle Company of Brantford also sold highwheelers with its name attached.
4 See Humber, *Freewheeling*, 49.
5 Several artisan-built machines survive, including a boneshaker in storage at Black Creek Pioneer Village in Toronto, a highwheeler in a barn at the Ontario Agricultural Museum in Milton, and another in a private collection near Goderich. A beautifully crafted oak boneshaker sold at auction in Niagara-on-the-Lake in 1996.
6 This machine seems to be based on the boneshaker 'serpentine' frame and has more in common with Lallement's velocipede than Michaux's design. Both wheels are the same size (they were probably taken off an agricultural machine), and the neck has supports attached to the backbone. This single-backbone machine must have been 'whippy as an eel' to ride.
7 See Humber, *Freewheeling*, 50. It seems likely that Fane (of Toronto) and the Goold Bicycle Company (of Brantford) began selling ordinary bicycles in 1888 and safety bicycles soon thereafter – in the latter case, several years earlier than the date listed in Appendix C of G. Donald Adams, *Collecting and Restoring Antique Bicycles.*
8 An 1892 catalogue of the Goold Bicycle Company of Brantford is in the collection of the Toronto Public Library's archives. Measuring 2 inches by 3 inches, with only a few line drawings, it is probably the first such list produced by this company. Just a few years later, Goold catalogues were larger and colourfully illustrated with a wider range of models. Their 'Red Bird' marque was not introduced until around 1895.
9 Merrill Denison, *CCM Good Sports: The Story of the First Fifty Years*, 22. Denison does not list a source for this statement.
10 Jacob Viner, *Dumping: A Problem in International Trade*, 192.
11 Robert Craig Brown, *Canada's National Policy, 1883–1900: A Study in Canadian-American Relations.*
12 In Canada, 'livery' meant a stable or storage area for horses or bicycles, which is quite different from the British use of the term.
13 It is not known which firm made them, but one of the bicycle manufacturers based in Toronto is the most likely possibility. The practice of attaching 'house names' continues to this day: major retailers frequently attach their own brand names to appliances made under contract by leading manufacturers.

14 Today, many small computer firms follow exactly the same strategy, purchasing all the necessary components and assembling them according to their own specifications.
15 The family's history can be found in Jane and Costas Varkaris, *The Pequegnat Story.*
16 These dates are listed by G. Donald Adams, *Collecting and Restoring Antique Bicycles*, 351.
17 The first known Canadian Racycle advertisement appeared in German in the *Berliner Journal* on 4 March 1897.
18 Varkaris, *The Pequegnat Story*, 18 claims that the Racycle crank hangar was invented principally by Arthur Pequegnat. This seems very unlikely since Racycle had been making bicycles with this patented crank hangar in the United States for a year before the Canadian Racycle was manufactured; in addition, the patent was not registered in Canada.
19 Humber (1986), 52.
20 Pope's contribution is assessed by Bruce Epperson, 'Failed Colossus: Albert A. Pope and the Pope Manufacturing Company, 1876–1900.' and Glen Norcliffe, 'Popeism and Fordism: Examining the Roots of Mass Production.' A case is made that Pope was a more important figure in the evolution of U.S. mass production than is suggested by David Hounshell, *From the American System to Mass Production,1800–1932: The Development of Manufacturing Technology in the United States*, chap.5; Hounshell sees Pope's contribution as more limited. For further insights into early mass production of bicycles see 'A Trip Through a Large Bicycle Factory' (1897) and André Vant, *L'Industrie du cycle dans la région stéphanoise.*
21 S.H. Oliver and D.H. Berkebile, *Wheels and Wheeling: The Smithsonian Cycle Collection.*
22 Hounshell, *From the American System to Mass Production.*
23 This was reported by Colonel Pope's great-grandson, in Albert Pope, 'Colonel Pope and the Founding of the U.S. Bicycle Industry.'
24 J.D. Womack, D.J. Jones, and D. Roos develop this argument in their 1990 book, *The Machine That Changed the World*, 26–7.
25 This is briefly reported in *Massey-Harris: A Historical Sketch, 1846–1926.*
26 Richard Harmond, 'Progress and Flight: An Interpretation of the American Cycle Craze of the 1890s,' 250–1
27 Denison, *CCM Good Sports*, 22, states that the Canadian Parliament adopted its first anti-dumping legislation in 1895. Viner states, however, that the first genuine anti-dumping clause in Canadian trade legislation can be found in the Customs Tariff Act of 1904; see Viner *Dumping.*
28 Details are given in 'CCM Prospectus,' 266, 278–9.
29 The history of this company is reported in Denison, *CCM Good Sports*. Unfortunately, no sources are cited in this book, so the accuracy of facts cannot be verified.
30 Denison, *CCM Good Sports*, 23. These are crude estimates – no accurate records are available.
31 David B. Perry, *Bike Cult: The Ultimate Guide to Human Powered Vehicles.*
32 Denison, *CCM Good Sports*, 32–4.

33 Jim Fitzpatrick, *The Bicycle and the Bush: Man and Machine in Rural Australia.*

Four: Bells and Whistles

1 According to *The Oxford Dictionary of New Words* (1991), the term can be traced to a report on microcomputers in the *Sunday Times*, 26 August 1984. It has since been widely used with respect to automobiles, although The *Longman Register of New Words* (1989) suggests that this application has a later date: see the *Daily Telegraph*, 26 December 1987.
2 Robert A. Smith, *The Social History of the Bicycle*, 24.
3 Such marketing strategies are also common in the tourist and personal computer industries. A ticket for a ship's cruise is often a loss leader: profits are made during the cruise on drinks, gambling, and day excursions at each port of call. Likewise, the purchase of a personal computer is often followed by the acquisition, over a long period, of additional hardware and software.
4 Nicholas Oddy, 'An Invaluable Refinement: The Aesthetic of the British Cycle Accessory in the Late 19th and Early 20th Centuries.'
5 Velocipede riders, by contrast, placed the arch of the foot against the pedal.
6 These historical details were gleaned from the *Globe and Mail*, 17 October 1996.
7 Oddy's typology was developed for a different purpose – to illuminate the aesthetic of the bicycle – but the generic types are relevant to the present discussion.
8 Frederick Alderson, *Bicycling: A History*, 74, lists the items of a bicyclist's correct outfit, with the approximate prices of each one.
9 Anita Rush, 'The Bicycle Boom of the Gay Nineties: A Reassessment,' 3. Some authors make excessively grand claims for the transformative social impact of the bicycle. Although it was important, it formed only part of a larger set of changes in women's roles, both political and social.
10 Small velocipedes designed for children, both bicycles and tricycles, were sold during the boneshaker phase, although sales in Canada were limited.
11 See Pryor Dodge, *The Bicycle*. Many of these illustrations are in Dodge's personal collection and have been displayed in an excellent exhibition of the bicycle, which has visited a number of North American cities.
12 See H.E. Stephenson and Carlton McNaught, *The Story of Advertising in Canada.*
13 These styles were also adopted by foreign manufacturers selling to the French market, mainly by commissioning Parisian artists. For example, in Dodge, *The Bicycle*, 160, there is a colour illustration of a 1900 Massey-Harris advertisement showing a woman with long tresses that curl round the handlebars of her bicycle, while she holds a silver maple leaf in her right hand. The image was drawn by a French poster artist, E. Célos, and printed by the Paris firm, Camille Sohet.
14 Stephenson and McNaught, *The Story of Advertising in Canada*, in their study of Canadian advertising, present a series of bicycle advertisements that are similarly restrained and informative.
15 The photographer, Frederick Doig, was born in Saint John in 1875 and died in 1949. In 1897 Doig won first prize in a photographic exhibition for his landscapes of Saint John.

16 Stephenson and McNaught, The Story of Advertising in Canada, 136.
17 Referred to as an economy of signs by Scott Lash and John Urry, Economies of Signs and Space.
18 Post-modernists would argue that, today, there is no single mainstream of cultural modernity, but a diversity of cultural narratives, each with its own logic. From a technological perspective, however, there is strong evidence of a cutting edge that attracts trendsetters who wish to be identified as such. Be it all-terrain vehicles, notebook computers, in-line skates, or cell phones, these trendsetters possess various products that act as signifiers of cultural modernity.
19 David B. Perry, *Bike Cult: The Ultimate Guide to Human Powered Vehicles.*
20 Perry, *Bike Cult*, 497.

Five: Bad Roads, Good Roads

1 This relationship, which was noted by Gershuny and Miles in their study of the *new service economy*, has formed a part of numerous product- and service-led waves. See Jonathan Gershuny and Ian Miles, *The New Service Economy*. They suggest, for example, that 'investment in major infrastructure ... made possible the diffusion of many of the new products of the post-war boom' (122).
2 Details are given in the CTC. *Monthly Gazette* 6, p.17. Both clubs donated £50 to launch the Association.
3 This lobby group, which became the Good Roads Association, was vigorously and widely promoted through the 1890s by Colonel Albert Pope, owner of the Pope Manufacturing Company. Pope's claim that he was performing a public service to improve roads by writing numerous letters to the press, lecturing across the country, subsidizing *Outing* magazine, and posting Good Roads pamphlets at his own expense is somewhat disingenuous. These activities gave Columbia bicycles excellent publicity and dramatically increased the road mileage suitable for bicycling. Sales of Columbia bicycles almost certainly increased as a result of these efforts. Details are given in many of the clippings in the surviving scrapbooks of Colonel Albert Pope (Connecticut Historical Society).
4 *Cycling*, 1, 20 (1892), 237.
5 Anita Rush, 'The Bicycle Boom in the Gay Nineties: A Reassessment,' 4.
6 Rush, 'The Bicycle Boom, 4 (emphasis added).
7 C.W. Hurndall, 'T.B.C., Last Run of the Season,' 6.
8 Brian Kinsman, *Around the World Awheel: The Adventures of Karl Creelman*, 43.
9 Macadam may be water bound, or tar bound. Based on his research into road technology, Nicholas Oddy suggests that these roads were probably water bound and rolled. I have not found confirmation, but light road rollers were advertised in *Good Roads* magazine by this date.
10 In figure 5.2 an engraving is shown of the Lorne Bridge built over the Grand River in Brantford, Ontario, in about 1880. This intriguing image, which appears in *Picturesque Canada*, is arguably the most anomalous picture in either of the two volumes: it celebrates the modern, as opposed to the picturesque, in a town that was noted for its industrial activity. Depicted are the new bridge, its new lighting, carriages, and an

ordinary bicycle crossing it. See George Munro Grant, ed., *Picturesque Canada*, 463.

11 Reported in *News Letter*, San Francisco, 12 March 1892, and found in Scrapbook 3, the Pope Scrapbooks (Connecticut Historical Society, Hartford Conn.).

12 *Cycling* 1, 3 (1890), 19.

13 Bagg's encyclopædism provides an interesting illustration of ideas on modernism and geographical classification of knowledge; see Charles W.J. Withers, 'Encyclopædism, Modernism, and Geographical Knowledge.'

14 Bagg was a small man, 5 feet 6 inches tall; hence, the small wheel.

15 Karl Kron, *Ten Thousand Miles on a Bicycle*, 285.

16 Kron, *Ten Thousand Miles*, 286.

17 Kron, *Ten Thousand Miles*, 290. As noted in chapter 1, this was not correct.

18 Kron, *Ten Thousand Miles*, 293.

19 Kron, *Ten Thousand Miles*, 310.

20 Kron, *Ten Thousand Miles*, 311.

21 Kron, *Ten Thousand Miles*, 314.

22 Today, when riding a century on an ordinary, a competent rider can expect to average ten miles per hour. Modern-day century rides usually start at 6 a.m., and even on hilly courses with regular breaks, most riders finish between 5 and 8 p.m. Riders in the 1880s were therefore riding about 25 per cent slower, owing to the bad roads and the frequent need to walk.

23 Kron, *Ten Thousand Miles*, 318.

24 Kron, *Ten Thousand Miles*, 325.

25 W.S. McKay, 'Highway Development in Ontario, 1793–1900,' 58.

26 The nature of statute labour is described in the December 1895 report of the Ontario Good Roads Association, Canadian Inventory of Historical Manuscripts, microfiche 63060, Ottawa.

27 Report of the Ontario Good Roads Association, December 1895, 4.

28 Reported in the *Good Roads Association Fiftieth Anniversary Brochure*, University of Toronto, Fisher Rare Books Collection, MS Collection 193 (Royal Canadian Institute) box 121, folder 12.

29 Reported by W.S. McKay, *Highway Development in Ontario*. The author was the son of the first treasurer of the Association, K.W. McKay.

30 Ontario Good Roads Association, minutes of meeting held in the city of Guelph, 11 December 1895, Canadian Inventory of Historical Manuscripts, microfiche 63060, 1.

31 McKay, *Highway Development in Ontario*, 76.

32 John C. Lehr and H. John Selwood, 'The Two-Wheeled Workhorse: The Bicycle as Personal and Commercial Transport in Winnipeg.'

33 Lehr and Selwood, *The Two-Wheeled Workhorse*.

34 I would argue that Rush's data prove the contrary. For example, she states that in 1897, only 120 of Calgary's population of 4000 were cyclists. Given the usual youthful population age distributions of that era and given the number of indigenous Canadians and impoverished immigrants then living in the city, bicycle ownership among males of Calgary's middle class, aged twenty to forty must have been about one in three.

The presence of ninety bicycle shops in Toronto in 1895, selling a total of 18,000 bicycles, is also evidence of large numbers of bicyclists, as is the proliferation of bicycle clubs such that Ottawa boasted six in 1896.

35 The *Revised Statutes of Ontario* of 1897 (Toronto: Queen's Printer) compiled all the acts that had been passed since 1887. The passages relating to bicycles are found in the Municipal Act, chapter 223, section 14. By part VII of this act municipalities were given various powers, including the regulation of bicycles and tricycles (which involved recognizing them as vehicles). The Amendment to the Municipal Act of 1897, 60V, c.45, s.19, permitted the setting apart of a fraction of a road to create a bicycle path and listed penalties for other vehicles that trespassed on this path; c.45, s.51, forbade bicycles from using sidewalks; c.56, s.1, permitted municipalities to pass by-laws controlling bicycles and tricycles. The Amendment to the Municipal Act of 1896, 59V, c.51, s.39, gave the Canadian Wheelmen's Association the right to erect and maintain, at the Association's own expense, signposts at intersections and elsewhere along roads (such as the signpost in figure 7.12).

36 This was reported in McKay, 'Highway Development in Ontario, 1793–1900,' a paper marking the fiftieth anniversary of the Ontario Good Roads Association. If the report is correct, the effort was unsuccessful, since there is no record in the *Canada Gazette* or the *Acts of the Parliament of Canada* for the years 1894–1900 of such an act's being introduced.

37 Moses's work is critically appraised by Marshall Berman, in part 5 of his *All That Is Solid Melts into Air.*

Six: The Cycling Crowd

1 Throughout this chapter, the term 'bicycle' should be taken, where appropriate, to include tricycles. This avoids having to repeat the phrase 'bicycles and tricycles' numerous times. The generic 'machine' and the participle 'cycling, are used from time to time as an alternative.

2 Robert A. Smith, *The Social History of the Bicycle: Its Early Life and Times in America*; and J.B. Townsend, 'The Social Side of Cycling.'

3 It is of interest that Bishop's contribution on the social and economic impacts of the bicycle was published in 1896, at the peak of the boom; see Joseph B. Bishop, 'Social and Economic Influence of the Bicycle.'

4 Heather Watts, *Silent Steeds: Cycling in Nova Scotia to 1900*, 33.

5 Anita Rush, 'The Bicycle Boom of the Gay Nineties: A Reassessment,' 1–3.

6 Smith, *The Social History of the Bicycle,* 13, 25.

7 One caveat should be noted: during the high-bicycle phase, older cyclists did potter around parks and do some touring on tricycles. In Britain, tricycles were also raced by younger riders, but details of any tricycle races being held in Canada have still to be found.

8 The term was also applied to a type of bicycle produced in the mid 1890s, sporty in appearance but not necessarily very fast or well made. Typically, it had drop handlebars, a sloping top-tube, and a short wheelbase.

9 *Cycling* 1, 12 (1892), 123.

10 Andrew Ritchie, 'The Hon. Keith-Falconer, G. Lacy Hillier, Class, and the Early Days of Bicycle Racing.'
11 In this respect, the bicycle differs from the automobile: continued innovation to 'top-end' cars, such as the BMW and Jaguar, have allowed them to remain positional goods over a long period. It could be argued that in the 1990s – with the advent of titanium, molybdenum, and carbon-fibre frames, suspension forks, and new gear-change systems – the mountain bicycle has become a focus of innovation and once again has made the bicycle a positional good.
12 In the twentieth century, French-Canadians have been in the forefront of Canadian cycle racing.
13 A similar picture emerges from group photographs of other clubs. Marvin Thomas has provided me with a picture of the Stratford Bicycle Club in 1886. Of the thirty-seven members identified, only one name does not appear to be of British origin.
14 *Cycling* 1, 2 (1890), 9.
15 *Cycling* 1, 1 (1890), 7.
16 *Cycling* 1, 2 (1890), 10.
17 Mr Miall was the son of an MP in the British Parliament and a businessman in Oshawa before joining the dominion public service. He rose to become commissioner of Inland Revenue, and commissioner of standards, a member of the royal commission to investigate the Canadian Pacific Railway, and a prominent figure in Ottawa society. His Ottawa address was Russell House.
18 There is no evidence that Canadian women rode boneshakers. See S. Michael Wells, 'Ordinary Women: High Wheeling Ladies in Nineteenth Century America' for an overview.
19 The exception to this rule was found in the circus and theatre, where the normal codes of Victorian conduct were waived. For instance, Mlle Louise Armaino of Montreal is depicted in the *Wheelmen's Gazette* of 1885 racing a high bicycle as part of what seems to be an 'entertainment.'
20 *Cycling* 1, 8 (1890), 60. This is an extraordinary twist of logic, since it was the prejudices of men, as much as women, that had for a generation kept women away from bicycles.
21 *Cycling* 1, 14 (1890), 150.
22 *Cycling* 7, 10 (1897), unpaginated.
23 Frances Willard, *A Wheel within a Wheel: How I Learned to Ride the Bicycle*. Following its publication, this book was said to be a best-seller. Many copies survive to this day.
24 Phillip Mackintosh, 'Wheel within a Wheel: Frances Willard and the Feminisation of the Bicycle.'
25 This position was echoed by R.L. Dickinson, 'Bicycling for Women: Some Hygienic Aspects of Wheeling.'
26 The Temperance Union named its headquarters for Frances Willard. A century later, Willard Hall still stands in downtown Toronto.
27 Mackintosh, 'Wheel within a Wheel,' 26.
28 Charles E. Pratt, *The American Bicycler: A Manual*.

29 See Ken Smith, ed., *The Canadian Bicycle Book.*
30 *Cyclist and Wheel World Annual,* 284.
31 Smith, *The Canadian Bicycle Book,* 8.
32 This ethos is evident in H. English, 'The Toronto Bicycle Club,' and in the minutes of the Montreal Bicycle Club for that period.
33 *Cycling* 2, 10 (1892), 135.
34 It was the working classes who rode on Sundays (their only day off work).
35 *Cycling* 1, 8 (1890) 73.
36 *Cycling* 2, 10 (1892), 138.
37 *Cycling* 2, 14 (1892) 240.
38 Christopher Armstrong and H.V. Nelles, *The Revenge of the Methodist Bicycle Company,* 157.
39 Armstrong and Nelles, *The Revenge,* 157, 167.
40 Armstrong and Nelles, *The Revenge,* 170.
41 John Wesley's *Deed of Declaration* (the basic text of Methodism), which was published in 1784, brought rational eighteenth-century Enlightenment thinking to bear upon Christian life. Christian duty lay in hard work, thrift, orderly methods, and entrepreneurialism. Success in business was a good thing, provided the fruits thereof were put to the common good.
42 Michael Bliss, *A Canadian Millionaire: The Life and Business Times of Sir Joseph Flavelle, bart., 1859–1939.*
43 Rush, *The Bicycle Boom of the Gay Nineties.*
44 A case can be made, however, that the advent of the safety bicycle allowed women to bring to an end the overtly masculinist phase of the high bicycle.
45 See Elizabeth Wilson, *Adorned in Dreams: Fashion and Modernity*; Elizabeth Wilson and Lou Taylor, *Through the Looking Glass: A History of Dress from 1860 to the Present Day*; and Susan D. Friedlander, 'Popularization of the Bicycle in the United States in the Late 1890s: Its Effect on Women's Dress and Social Codes.' For contemporary comments on women and cycling see R.L. Dickinson, 'Bicycling for Women: The Puzzling Question of Costume'; and Henry J. Garrigues, 'Woman and the Bicycle.'
46 The choice of Sparks Street is curious. Individual and club bicycle rides were usually made into the countryside. For the recreational cyclist, a downtown street was not a common destination.
47 In economic terms, a technology rent is a form of economic rent, a payment for a scarce resource that is in demand.
48 The product cycle is generally seen to pass through four stages: (1) an early phase in which production is in very small numbers, owing to frequent innovation and a small market; (2) growth of production runs as demand grows, although constant technological changes require frequent retooling; competition between firms is over technology rather than price; (3) increasing standardization of components and parts allowing mass production; prices start to drop; (4) virtual cessation of innovation, so that long production runs are possible; competition between firms is over price rather than technology; factories decentralize to low-wage areas where bicycles can be made as cheaply as possible. Discussion of the product cycle can be found in

R.A.Erickson, 'The Filtering Down Process: Industrial Location in a Non-metropolitan Area'; and J.M.Utterback, 'The Dynamics of Product and Process Innovation in Industry.'

Seven: Larger Spaces and Visible Places

1 This information was gleaned from an article published in the *University of Toronto Magazine* 24, 1 (Autumn 1996) 27, and from a personal interview with Dr Luke Irwin of Orillia, grandson of the pioneer cyclist.
2 Details of this litigation are given in *Cycling* 1, 24 (11 November 1891), 287.
3 It was not uncommon for members of bicycle clubs to be involved in litigation with obstructive teamsters. For example, in the summer of 1890 Mr English of the Toronto Bicycle Club was run down by Mr McCuaig, the driver of an express wagon. His bicycle was demolished. English sued and received $100 damages, plus costs (*Cycling* 1, 1 [November 1890], 4). In an act of solidarity, the Canadian Wheelmen's Association had offered to pick up English's legal costs.
4 *Cycling* 2, 10 (14 April 1892), 140.
5 Edith G. Firth, *Toronto in Art*, 88. The accompanying painting (89), entitled *The Legislative Building, seen from University Avenue*, 1897, shows a narrow leafy avenue with eleven cyclists standing chatting or riding on the sidewalks.
6 Pierre Berton, *Klondike: The Last Great Gold Rush*, 117–20.
7 Pierre Berton, *The Klondike Quest: A Photographic Essay.*
8 The drop-handlebar sports bicycle on the right would not be at all practical in the Yukon.
9 William Humber, *Freewheeling: The Story of Bicycling in Canada*, 45–7.
10 An amusing anecdote concerning Klondike bicycles occurred a few years later. The local hockey team, the Dawson City Nuggets, were challengers for the Stanley Cup in the winter of 1905/06. This was a winter without snow, however, and on 18 December 1905 the members of the team left Dawson City bound for Ottawa via Whitehorse on bicycles. They switched to sleighs further on when snow fell, and twenty-eight days later reached their final destination, where they played the Ottawa Silver Seven, and lost badly.
11 Rob Shields, *Places on the Margin.*
12 The Half Way House still stands, but not at its original site. It was trucked in the 1960s to Black Creek Pioneer Village, where it stands bereft of its original place meaning.
13 This tour began in Portland, Maine, and went to Saint John, Grand Falls, Andover, Florenceville, and Fredericton in New Brunswick. Figures 1.3 and 6.2 are photographs taken of this tour.
14 The five episodes, all published in the *Canadian Magazine* under the headline 'A Canadian Bicycle in Europe,' were as follows: chap.1 'From Toronto to Gibraltar,' 6, 6 (April 1896), 530–6; chap. 2 'Gibraltar and Algiers,' 7, 1 (May 1896), pp.11–17; chap. 3 'La Bella Napoli,' 7, 2 (June 1896), 111–20; chap. 4 'Vesuvius and Pompeii,' 7, 3 (July 1896), 216–26; and chap. 5 'Rome, Florence, Venice,' 7, 4 (August 1896), 323–32.

15 *Canadian Magazine* 7, 4 (August 1896), 327.
16 Brian Kinsman, *Around the World Awheel: The Adventures of Karl Creelman.*
17 David Harvey, *The Condition of Postmodernity.*
18 James M. Cameron in *The Canadian Beaver Book: Fact, Fiction and Fantasy* shows that the beaver has served as Canada's most enduring national symbol.
19 John Berger, 'Understanding a Photograph,' provides valuable insights into the meaning of photographs.
20 Frederick Alderson, *Bicycling: A History*, 65.
21 Joan M. Schwartz, 'The Prairie, on the Banks of the Red River, Looking South: More Than "Competent Description of an Intractably Empty Landscape,"' 3.
22 Charles Baudelaire, 'Le peintre de la vie moderne.'
23 See several of the essays in Keith Tester, ed., *The Flâneur.*
24 Norcliffe, 'Out for a Spin: The Flâneur on Wheels.'
25 Tester, *The Flâneur.*
26 H.P. Maxim, *Horseless Carriage Days*, 4–5.
27 David Hounshell, *From the American System to Mass Production*, 214.

Eight: Pedaller's Progress

1 Nigel Thrift, 'Shut up and Dance, or Is the World Economy Knowable?' 15.
2 See Jean Baudrillard, 'Modernity.'
3 The Islamic religion, for example, has taken on very distinct characteristics in different countries.
4 Many insights into this transformation can be found in Derek Gregory, *Geographical Imaginations.*
5 Baudrillard, 'Modernity,' 68–69.
6 P. Anderson, 'Modernity and Revolution.'
7 Philip Cooke, 'Modernity, Postmodernity and the City,' 475.
8 Baudrillard, 'Modernity,' 63.
9 Baudrillard, 'Modernity,' 63.
10 Paul Rabinow, *French Modern*, 139.
11 J. Habermas, *The Philosophical Discourse of Modernity*.See also John Law, *Organizing Modernity* and Timothy W. Luke, *Social Theory and Modernity.*
12 Krsihan Kumar, *From Post-Industrial to Post-Modern Society*, 83.
13 The significance of the Viet Nam War is that many postmodernists see it as a divide, marking the beginning of a new era when the canons of modernity have been under intense scrutiny and when several alternative cultures have gained acceptance. Mike Featherstone, *Consumer Culture and Postmodernism*, examines this cultural fragmentation and the collapse of the boundaries between culture and everyday life. The nature of this cultural sea change is examined by Anthony Giddens, 'Modernism and Post-modernism,' and Patrick Waugh, *Postmodernism: A Reader.* One consequence of the rise of postmodern cultural values is that old bicycles have become a part of the 'heritage industry'; see Robert Hewison, *The Heritage Industry.*
14 Malcolm Bradbury and James McFarlane, *Modernism, 1890–1930.*

Select Bibliography

Adams, G. Donald. *Collecting and Restoring Antique Bicycles*. 2nd ed. Orchard Park, N.Y.: Pedaling History – Burgwardt Bicycle Museum, 1996.

Alderson, Frederick. *Bicycling: A History*. New York: Praeger, 1972.

Allen, G.C. *The Industrial Development of Birmingham and the Black Country, 1860–1927*. New York: Augustus M. Kelley, 1966.

Anderson, P. 'Modernity and Revolution.' *New Left Review*, no. 144 (1984), 96–113.

Armstrong, Christopher, and H.V. Nelles. *The Revenge of the Methodist Bicycle Company: Sunday Streetcars and Municipal Reform in Toronto, 1888–1897*. Toronto: Peter Martin, 1977.

Babaian, Sharon. *The Most Benevolent Machine: An Historical Assessment of Cycles in Canada*. Ottawa: National Museum of Science and Technology, 1998.

Bartleet, H.W. *Bartleet's Bicycle Book*. London: E.J. Burrow, 1931.

Baudelaire, Charles. 'Le peintre de la vie moderne,' *Oeuvres Complètes*. Paris: La Pléiade Editions NRF, 1861.

Baudrillard, Jean. 'Modernity,' *Canadian Journal of Political and Social Theory* 11, 3 (1987), 63–72.

Baum, Gregory. 'Modernity: A Sociological Perspective.' In *The Debate on Modernity*. Ed. Claude Geffré and Jean-Pierre Jossua. London: SCM Press, 1992.

Bauman, Zygmunt. *Modernity and Ambivalence*. London: Polity Press, 1991.

Berger, John. 'Understanding a Photograph.' In *Classic Essays on Photography*. Ed. Alan Trachtenberg. New Haven, Conn.: Leete's Books, 1980.

Berman, Marshall. *All That Is Solid Melts into Air*. New York: Penguin Books, 1988.

Berton, Pierre. *Klondike: The Last Great Gold Rush, 1896–1899*. Toronto: McClelland and Stewart, 1972.

– *The Klondike Quest: A Photographic Essay*. Boston: Little, Brown, 1983.

Bijker, Weibe E. *Of Bicycles, Bakelites and Bulbs: Towards a Theory of Socio-Technical Change*. Cambridge, Mass.: MIT Press, 1995.

Bishop, Joseph B. 'Social and Economic Influence of the Bicycle.' *Forum and Century* 21 (1896), 680–9.

Bliss, Michael. *A Canadian Millionaire: The Life and Business Times of Sir Joseph Flavelle, Bart., 1859–1939.* Toronto: Macmillan, 1978.

Bolger, William C. *Smithville: The Result of Enterprise.* Mount Holly, N.J.: Burlington County Cultural and Heritage Commission, 1980.

Boulton, Constance Rudyard. 'A Canadian Bicycle in Europe.' *The Canadian Magazine* 6, 6 (1896), 530–6; 7, 1 (1896), 11–17; 7, 2(1896) 111–20; 7, 3 (1896), 216–26; 7, 4 (1896), 323–32.

Bowerman, Les. 'Lewis Gompertz and His Addition to the Velocipede.' *Proceedings of the Third International Conference of Bicycling History.* Neckarsulm, Germany: Deutsches Zweirad-Museum, 1992.

Bradbury, Malcolm, and James McFarlane. *Modernism, 1890–1930.* Harmondsworth, Middlesex: Penguin, 1976.

Brown, J.J. *Ideas in Exile.* Toronto: McClelland and Stewart, 1965.

Brown, Robert Craig. *Canada's National Policy 1883–1900: A Study in Canadian-American Relations.* Princeton, N.J.: Princeton University Press, 1964.

Cameron, James M. *The Canadian Beaver Book: Fact, Fiction and Fantasy.* Burnstown, Ontario: General Store Publishing House, 1991.

'CCM Prospectus.' *Monetary Times*, 1 September, 1899, 278–9.

Clark, T.J. *The Painting of Modern Life: Paris in the Art of Manet and His Followers.* London: Thames and Hudson, 1985.

Clayton, Nick. 'The Development of the Suspension Wheel,' *Proceedings of the Second International Conference on Cycle History.* St Etienne: Le Musée d'Art et d'Industrie de Saint-Etienne, 1991.

– *Early Bicycles.* Princes Risborough, Bucks: Shire Publications, 1986.

– 'Who Invented the Penny-Farthing?' In *Cycle History: Proceedings of the 7th Cycle History Conference.* Ed. Rob van der Plas, San Francisco: Bicycle Books, 1997.

Cole, Terrence, ed. *Wheels on Ice.* Anchorage: Alaska Northwest, 1985.

Compagnon, Antoine. *The Five Paradoxes of Modernity.* Trans. Franklin Philip. New York: Columbia University Press, 1994.

Cooke, Philip. *Back to the Future: Modernity, Postmodernity and Locality.* London: Unwin Hyman, 1990.'

– 'Modernity, Postmodernity and the City,' *Theory, Culture and Society* 5 (1988), 475–92.

*Cyclist and Wheel World Annua*l. N.p, 1882.

Denison, Merrill. *CCM Good Sports: The Story of the First Fifty Years.* Toronto: CCM, 1946.

Dickinson, R.L. 'Bicycling for Women: Some Hygienic Aspects of Wheeling.' *Outlook*, 28 March 1896, 550–3.

– 'Bicycling for Women: The Puzzling Question of Costume.' *Outlook*, 25 April 1896, 751–2.

Dodds, Alastair. 'Kirkpatrick Macmillan, Inventor of the Bicycle: Fact or Hearsay?' *Proceedings of the Third International Conference of Bicycling History.* Neckarsulm,

Germany: Deutsches Zweirad-Museum, 1992.
Dodge, Pryor. *The Bicycle*. Paris, New York: Flammarion, 1996.
Dunford, Mick, and Diane Perrons. *The Arena of Capital*. London: Macmillan, 1983.
Dunham, Norman Leslie. 'The Bicycle Era in American History.' PhD thesis, Harvard University, 1956.
English, H. 'The Toronto Bicycle Club.' *Outing* 16 (1890), unpaginated, 2 pages.
Entrikin, J. Nicholas. *The Betweenness of Place: Towards a Geography of Modernity*. London: Macmillan, 1991.
Epperson, Bruce. 'Failed Colossus: Albert A. Pope and the Pope Manufacturing Company, 1876–1900,' *Cycle History: Proceedings of the 9th International Cycle History Conference*. Ed. Glen Norcliffe and Rob van der Plas. San Francisco: Bicycle Books, 1999.
Erickson, R.A. 'The Filtering Down Process: Industrial Location in a Non-metropolitan Area.' *Professional Geographer* 28 (1976), 254–60.
Featherstone, Mike. *Consumer Culture and Postmodernism*. London: Sage, 1991.
Firth, Edith G. *Toronto in Art: 150 Years through Artists' Eyes*. Toronto: Fitzhenry & Whiteside, 1983.
Fitzpatrick, Jim. *The Bicycle and the Bush: Man and Machine in Rural Australia*. Oxford: Oxford University Press, 1980.
Flink, James, J. *The Automobile Age*. Cambridge, Mass.: MIT Press, 1988.
Friedlander, Susan D. 'Popularization of the Bicycle in the United States in the Late 1890s: Its Effect on Women's Dress and Social Codes.' *Wheelman* 45 (1994) 2–11.
Garrigues, Henry J. 'Woman and the Bicycle,' Forum January 1896, 576–87.
Gershuny, Jonathan, and Ian Miles. *The New Service Economy*. London: Frances Pinter, 1983.
Giddens, Anthony. 'Modernism and post-modernism.' *New German Critique* 22 (1981), 15–18.
Glennie, Paul, and Nigel Thrift. 'Consumers, Identities, and Consumption Spaces in Early-Modern England.' *Environment and Planning A* 28 (1996), 25–45.
– 'Modernity, Urbanism, and Modern Consumption,' *Environment and Planning D: Society and Space* 10 (1992), 423–43.
Gougaud, M.B. 'Note sur Barthélémy Thimonnier.' *Proceedings of the Second International Conference on Cycle History*. St Etienne: Le Musée d'Art et d'Industrie de Saint-Etienne, 1991.
Granovetter, Mark. 'Economic Action and Social Structure: The Problem of Embeddedness.' *American Journal of Sociology* 91 (1985), 481–510.
Grant, George Munro, ed. *Picturesque Canada*. 2 vols. Toronto: Belden Press, 1882.
Gregory, Derek. *Geographical Imaginations*. Oxford: Blackwell, 1994.
– 'Interventions in the Historical Geography of Modernity: Social Theory, Spatiality and the Politics of Representation.' *Geografiska Annaler* 73B (1991), 17–44.
– 'Modernity.' *Dictionary of Human Geography*. 3rd ed. Ed. R.J. Johnston, Derek Gregory, and David M. Smith. Oxford: Blackwell, 1994.
Habermas, J. *The Philosophical Discourse of Modernity*. Cambridge: Polity Press, 1987.
Hall, Peter, and Paschal Preston. *The Carrier Wave: New Information Technology and the*

Geography of Innovation, 1846–2003. London: Unwin Hyman, 1988.
Harmond, Richard. 'Progress and Flight: An Interpretation of the American Cycle Craze of the 1890s.' *Journal of Social History* 5 (1971), 235–57.
Harris, R.C. 'Power, Modernity, and Historical Geography.' *Annals of the Association of American Geographers* 81 (1988), 671–83.
Harrison, A.E. 'The Competitiveness of the British Cycle Industry, 1890–1914.' *Economic History Review* 22 (1969), 287–303.
Harvey, David. *The Condition of Postmodernity*. Oxford: Blackwell, 1989.
Herlihy, David, V. 'H. Cadot: One of the Earliest Bicycle Makers in France.' *Cycle History: Proceedings of the 7th International Cycle History Conference*. Ed. Rob van der Plas. San Francisco: Bicycle Books, 1997.
– 'Lallement vs. Michaux: Who Truly Invented the Bicycle?' *Wheelman* 42 (1993), 2–16.
Herman, Edward S., and Noam Chomsky. *Manufacturing Consent: The Political Economy of the Mass Media*. New York: Pantheon Press, 1988.
Hewison, Robert. *The Heritage Industry*. London: Methuen, 1987.
Hillier, G. Lacy. 'Cycles Past and Present.' *Transactions of the Royal Scottish Society of Arts* 13 (1892), 243–57.
Hounshell, David. *From the American System to Mass Production, 1800–1932: The Development of Manufacturing Technology in the United States*. Baltimore: Johns Hopkins University Press, 1984.
Humber, William. *Freewheeling: The Story of Bicycling in Canada*. Erin, Ont.: Boston Mills Press, 1986.
Hurndall, C.W. 'T.B.C., Last Run of the Season.' *Cycling* 1, 1 (1890), 6.
Jackson, Peter. 'Constructions of Culture, Representations of Race: Edward Curtis's "Way of Seeing."' In *Inventing Places: Studies in Cultural Geography*. Ed. Kay Anderson and Fay Gale. Melbourne: Longman Cheshire, 1992.
Johnston, R.J., Derek Gregory, and David M. Smith. *Dictionary of Human Geography*. 3rd ed. Oxford: Blackwell, 1994.
Kinsman, Brian. *Around the World Awheel: The Adventures of Karl Creelman*. Hantsport, N.S.: Lancelot Press, 1993.
Kobayashi, Keizo. 'Les Frères Olivier.' *Proceedings of the Second International Conference on Cycle History*. St Etienne: Le Musée d'Art et d'Industrie de Saint-Etienne, 1991.
– *Histoire du vélocipède de Drais à Michaux: Mythes et réalités*. Paris: l'École Pratique des Hautes Études, 1993.
– 'The Velocipede in France, from Karl von Drais to Pierre Michaux, 1818–1861.' *Proceedings of the Third International Conference of Bicycling History*. Neckarsulm, Germany: Deutsches Zweirad-Museum, 1992.
Kron, Karl. *Ten Thousand Miles on a Bicycle*. New York: Karl Kron, 1887.
Kumar, Krishan. *From Post-Industrial to Post-Modern Society*. Oxford: Blackwell, 1995.
Laget, Françoise, and Serge Laget. *Le Cyclisme*. Courlay: Jadault et fils, 1978.
Lash, Scott, and John Urry. *Economies of Signs and Space*. London: Sage, 1994.
Law, John. *Organizing Modernity*. Oxford: Blackwell, 1994.
Lefebvre, Henri. *The Production of Space*. Oxford: Blackwell, 1991.

Lehr, John C., and H. John Selwood. 'The Two-Wheeled Workhorse: The Bicycle as Personal and Commercial Transport in Winnipeg.' *Urban History Review* 28, 1 (1999), 3–13.

Leitch, Adelaide. *Floodtides of Fortune: The Story of Stratford and the Progress of the City through Two Centuries*. Stratford, Ont.: Corporation of the City of Stratford, 1980.

Lessing, Hans-Erhard. 'Around Michaux: Myths and Realities.' *Proceedings of the Second International Conference on Cycle History*. Ed. Brian Hayward and Andrew Millward. Cheltenham: Quorum Technical Services, 1995.

– *Karl von Drais: der Empire-Technologe Wird Rehabiliert*. Mannheimer Geschichtsblätter Neue Folge. Mannheim: Sigmaringen, 1996.

– 'Karl von Drais' Two-Wheeler – What We Know.' In *Proceedings of the First International Conference of Cycling History*. Museum of Transport, 1990.

Luke, Timothy W. *Social Theory and Modernity*. Newbury Park, Cal.: Sage, 1990.

Mackintosh, Phillip. '"Wheel within a Wheel": Frances Willard and the Feminization of the Bicycle.' In *Cycle History: Proceedings of the 9th International Cycle History Conference*. Ed. Glen Norcliffe and Rob van der Plas. San Francisco: Bicycle Books, 1999.

Massey-Harris: A Historical Sketch, 1846–1926. Toronto: Massey-Harris Company, 1926.

Maxim, H.P. *Horseless Carriage Days*. New York: Harper, 1937.

McConagle, Seamus. *The Bicycle in Life, Love, War and Literature*. London: Pelham Books, 1968.

McKay, W.S. 'Highway Development in Ontario, 1793–1900.' *Municipal World*, March 1944. University of Toronto, Fisher Rare Books Collection. MS Collection 193 (Royal Canadian Institute). Box 121. Folder 12. p. 58.

Millward, Andrew. 'The Cycle Trade in Birmingham, 1890–1920.' *Proceedings of the Third International Conference of Bicycling History*. Neckarsulm, Germany: Deutsches Zweirad-Museum, 1992.

– 'The Genesis of the British Cycle Industry, 1867–1872.' *Proceedings of the First International Conference of Cycling History*. Glasgow: Museum of Transport, 1990.

Naremore, James, and Patrick Brantlinger, eds. *Modernity and Mass Culture*. Bloomington: Indiana University Press, 1991.

Norcliffe, Glen. 'Colonel Albert Pope: His Contribution to Bicycle Manufacture and the Development of Mass Production.' In *Cycle History: Proceedings of the 7th International Cycle History Conference*. Ed. Rob van der Plas. San Francisco: Bicycle Books, 1997.

– 'Embedded Innovation: Canadian Cycle-Related Patents, 1868–1900.' In *Cycle History: Proceedings of the 9th International Cycle History Conference*. Ed. Glen Norcliffe and Rob van der Plas. San Francisco: Bicycle Books, 1999.

– 'Out for a Spin: The Flâneur on Wheels.' In *Cycle History: Proceedings of the 8th International Cycle History Conference*. San Francisco: Bicycle Books, 1998.

– 'Popeism and Fordism: Examining the Roots of Mass Production.' *Regional Studies* 31 (1997), 267–80.

Oddy, Nicholas. 'The Bicycle – an Exercise in Gendered Design.' In *Cycle History: Proceedings of the 5th International Cycle History Conference*. Ed. Rob van der Plas. San Francisco: Bicycle Books, 1995.

– 'An Invaluable Refinement: The Aesthetic of the British Cycle Accessory in the Late 19th and Early 20th Centuries.' In *Cycle History: Proceedings of the 7th International Cycle History Conference.* Ed. Rob van der Plas. San Francisco: Bicycle Books, 1997.

– 'Kirkpatrick Macmillan, the Inventor of the Pedal Cycle or the Invention of Cycle History.' *Proceedings of the First International Conference of Cycling History.* Glasgow: Museum of Transport, 1990.

Ogborn, Miles. *Spaces of Modernity: London's Geographies, 1680–1780.* London: Guildford, 1998.

Oliver, S.H., and D.H. Berkebile. *Wheels and Wheeling: The Smithsonian Cycle Collection.* Washington, DC: Smithsonian Institution Press, 1974.

'The Pedestrian's Accelerator.' *The Imperial Magazine or Compendium of Religious, Moral and Philosophical Knowledge* 1, 2 (April 1819), 143–4.

Perry, David B. *Bike Cult: The Ultimate Guide to Human Powered Vehicles.* New York: Four Walls Eight Windows, 1995.

Pietsch, Jost. 'The Karl Kech Bicycle – A Reassessment.' *Boneshaker* 15, 143 (1997), 4–5.

Pinch, Trevor J., and Weibe E. Bijker. 'The Social Construction of Facts and Artefacts: Or How the Sociology of Science and the Sociology of Technology Might Benefit Each Other.' In *The Social Construction of Technological Systems.* Ed. Weibe E. Bijker, Thomas P. Hughes, and Trevor J. Pinch. Cambridge Mass.: MIT Press, 1987.

Polanyi, Karl. 'The Economy as Instituted Process.' In *Trade and Market in the Early Empire.* Ed. Karl Polanyi et al. New York: Free Press, 1955.

Pope, Albert A. 'Colonel Pope and the Founding of the U.S. Bicycle Industry.' In *Cycle History: Proceedings of the 5th International Cycle History Conference.* Ed. Rob van der Plas. San Francisco: Bicycle Books, 1995.

Pratt, Charles E. *The American Bicycler: A Manual.* Boston: Rockwell and Churchill, 1880.

Precious, Carole. *Thomas Carbide Willson.* Don Mills, Ont.: Fitzhenry & Whiteside, 1980.

Pred, Allan, and Michael J.Watts. *Reworking Modernity: Capitalism and Symbolic Discontent.* New Brunswick, N.J.: Rutgers University Press, 1992.

Rabinow, Paul. *French Modern: the Norms and Forms of the Social Environment.* Cambridge, Mass.: MIT Press, 1989.

Ritchie, Andrew. 'The Hon. Keith-Falconer, G. Lacy Hillier, Class, and the Early Days of Bicycle Racing.' *Boneshaker* 15, 141 (1996), 26–34.

– *King of the Road: An Illustrated History of Cycling.* Berkeley, Cal.: Ten Speed Press, 1975.

Roberts, Derek. *Cycling History: Myths and Queries.* Birmingham: Pinkerton Press, 1991.

– *The Invention of Bicycles and Motorcycles.* London: Usborne, 1975.

– *The Invention of the Safety Bicycle.* 3rd ed. Mitcham, Surrey: private publication, 1993.

Rush, Anita. 'The Bicycle Boom of the Gay Nineties: A Reassessment.' *Material History Review* 18 (1983), 1–12.

Sack, Robert David. *Place, Modernity, and the Consumer's World.* Baltimore: Johns Hopkins University Press, 1992.

Sauvaget, Roland. 'Michaux/Lallement.' *Boneshaker* 14, 137 (1995), 12–15.
– 'Michaux-v-Lallement: The Conclusion?' *Boneshaker* 16, 152 (2000), 4–10.
Schofield, Robert E. *The Lunar Society of Birmingham: A Social History of Provincial Science and Industry in Eighteenth-Century England.* Oxford: Clarendon Press, 1963.
Schwartz, Joan M. 'The Prairie, on the Banks of the Red River, Looking South: More Than "Competent Description of an Intractably Empty Landscape."'Unpublished manuscript. 1996. National Archives of Canada, Ottawa.
Sharp, Archibald. *Bicycles and Tricycles: An Elementary Treatise on their Design and Construction.* 1896. Reprint Cambridge, Mass.: MIT Press, 1993.
Shields, Rob. *Places on the Margin: Alternative Geographies of Modernity.* London: Routledge, 1991.
Sissons, Constance Kerr. *John Kerr.* Toronto: Oxford University Press, 1946.
Smith, Ken, ed. *The Canadian Bicycle Book.* Toronto: D.C. Heath, 1972.
Smith, Robert A. *The Social History of the Bicycle: Its Early Life and Times in America.* New York: American Heritage Press, 1972.
Stephenson, H.E., and Carlton McNaught. *The Story of Advertising in Canada.* Toronto: Ryerson Press, 1940.
Street, Roger. *The Pedestrian Hobby-Horse: At the Dawn of Cycling.* Christchurch, Dorset: Artesius Publications, 1998.
Tester, Keith, ed. *The Flâneur.* London: Routledge, 1994.
Thompson, Christopher S. 'The Third Republic on Wheels: A Social, Cultural and Political History of Bicycling in France from the Nineteenth Century to World War II.' PhD dissertation, New York University, 1997.
Thrift, Nigel. 'Shut up and Dance, or Is the World Economy Knowable?' In *The Global Economy in Transition.* Ed. Peter Daniels and William Lever. Harlow, Essex: Addison Wesley Longman, 1997.
Townsend, J.B. 'The Social Side of Cycling.' *Scribner's Magazine* (June 1895), 704–8.
'A Trip through a Large Bicycle Factory.' *Scientific American* 77, 19 (November 1897), 292–3.
Utterback, J.M. 'The Dynamics of Product and Process Innovation in Industry.' In *Technological Innovation for a Dynamic Economy.* Ed. C.T. Hill and J.M. Utterback. New York: Pergamon Press, 1979.
Vant, André. 'L'industrie du cycle dans la région stéphanoise.' *Revue de géographie de Lyon* 49 (1974),155–84.
– *L'industrie du cycle dans la région stéphanoise.* Lyon: Éditions lyonnaise d'Art et d'Histoire, 1993.
Vant, André, and Jacqueline Dupuis. 'L'industrie stéphanoise du cycle ou la fin d'un système industriel localisé.' *Revue de géographie de Lyon* 68 (1993), 5–16.
Varkaris, Jane, and Costas Varkaris. The Pequegnat Story. Dubuque, Iowa: Kendall/Hunt, 1982.
'Velox.' *Velocipedes, Bicycles and Tricycles: How to Make and How to Use Them.* London: Routledge, 1869.
Viner, Jacob. *Dumping: A Problem in International Trade.* New York: A.M. Kelley, 1966.
Walden, Keith. *Becoming Modern in Toronto.* Toronto: University of Toronto Press, 1997.
Ward, David. 'A Comparative Historical Geography of Streetcar Suburbs in Boston,

Massachusetts and Leeds, England: 1850–1920.' *Annals of the Association of American Geographers* 54 (1964), 477–89.
Warner, Sam Bass, Jr. *Streetcar Suburbs: The Process of Urban Growth in Boston, 1870–1900.* Cambridge, Mass.: Harvard University Press, 1978.
Watts, Heather. *Silent Steeds: Cycling in Nova Scotia to 1900.* Halifax: Nova Scotia Museum, 1985.
Waugh, Patrick. *Postmodernism: A Reader.* London: Edward Arnold, 1992.
Weber, Eugen. *My France: Politics, Culture, Myth.* Cambridge, Mass.: Belknap Press, 1991.
– *Peasants into Frenchmen: The Modernization of Rural France, 1870–1914.* Stanford, Cal.: Stanford University Press, 1976.
Weber, Max. *The Protestant Ethic and the Spirit of Capitalism.* Trans. Talcott Parsons. 1904. Reprint, New York: Scribner, 1930 .
Wells, S. Michael. 'Ordinary Women: High Wheeling Ladies in Nineteenth Century America.' *Wheelman* 43 (1993), 2–14.
Willard, Frances. *A Wheel within a Wheel: How I Learned to Ride the Bicycle.* New York: Fleming H. Revell, 1895.
Wilson, Elizabeth. *Adorned in Dreams: Fashion and Modernity.* London: Virago, 1985.
Wilson, Elizabeth, and Lou Taylor. *Through the Looking Glass: A History of Dress from 1860 to the Present Day.* London: BBC Books, 1989.
Wilton, Elizabeth. 'Cloud Bound: The Western Landscapes of Marmaduke Matthews.' In *A Few Acres of Snow.* Ed. Paul Simpson-Housley and Glen Norcliffe. Toronto: Dundurn Press, 1992.
Winder, Gordon. 'The North American Manufacturing Belt in 1880: A Cluster of Regional Industrial Systems or One Large District?' *Economic Geography* 75 (1999), 71–92.
Withers, Charles W.J. 'Encyclopaedism, Modernism and the Classification of Geographical Knowledge.' *Transactions of the Institute of British Geographers* NS 21 (1996), 275–98.
Womack, J.D., D.J. Jones, and D. Roos. *The Machine That Changed the World.* Toronto: Collier-Macmillan, 1990.

Illustration Credits

Author's collection: Figure 2.1; Figure 2.7 (compiled by author); Figure 3.2 (*Cycling* 1, 7 [February 1891], 54); Figure 3.12; Figure 3.13 (*Cycling*, 27 April 1901); Figure 3.14 (*Cycling*, 18 May 1901); Figure 3.15 (*Cycling*, 25 May 1901); Figure 4.2; Figure 4.13 (*Cycling* 7, 10 [8 April 1897], n.p.); Figure 4.14 (*Cycling* 2, 5 [28 January 1892], 64); Figure 4.16; Figure 5.2 (George Munro Grant, ed., *Picturesque Canada*, Vol. 2 (Toronto: Belden Press 1882), 463); Figure 5.7 (compiled by author); Figure 5.9 (compiled by author); Figure 6.8 (redrawn by Hilary Norcliffe)
Bruce County Archives: Figure 7.2 (A970.15.2, neg.1917/87/#22, Box 9D); Figure 7.11 (A962.13.3, neg.A33/83-35, Box 7)
Canadian Intellectual Property Office (Patents Branch, Ottawa: various years): Figure 2.8; Figure 2.10
City of Collingwood Museum: Figure 6.1 (NO2407)
City of Vancouver Archives: Figure 7.3 (BU.P.311.N699)
Hamilton Public Library: Figure 2.4
Brian Kinsman: Figure 7.13
Ronald Miller: Figure 2.9; Figure 3.3; Figure 4.6; Figure 5.11; Figure 6.10
National Archives of Canada, Ottawa: Figure 1.6 (C76519); Figure 1.10 (C39096); Figure 2.3 (PA24047); Figure 2.5 (PA32392, Topley Collection); Figure 2.6 (PA196537); Figure 4.3 (C68844, McCurry Collection); Figure 5.3 (PA196533, George Schofield Collection); Figure 5.4 (PA196530, George Schofield Collection); Figure 5.5 (PA17208); Figure 5.10 (Map Division); Figure 6.4 (PA132226); Figure 6.9 (C79291, McCurry Collection); Figure 7.1 (PA 117883); Figure 7.7 (PA196536); Figure 7.8 (PA196353, Heneker Family Collection); Figure 7.12 (PA196531, George Schofield Collection); Figure 7.14 (PA16077); Figure 7.16 (PA32543, Topley Collection); Figure 7.18 (PA57608)
New Brunswick Museum, Saint John: Figure 4.15 (989.181.8, Frederick Doig Collection)

Notman Photographic Archives, McCord Museum, Montreal: Figure 1.5 (MP076/77(96); Roper Donation); Figure 1.11 (115,992-II); Figure 2.2 (78664); Figure 4.1 (78561-BII); Figure 4.4 (78577-II); Figure 4.5 (96758-BII); Figure 4.7 (116293); Figure 4.8 (112408-BII); Figure 4.9 (120688-BII); Figure 4.10 (120688-BII); Figure 4.11 (78679-B11); Figure 4.12 (106548-BII); Figure 5.13 (MP022/79(168)); Figure 6.5 (26273); Figure 6.7 (78594-BII); Figure 7.6 (MP2360(1)); Figure 7.10 (MP003/87(2)); Figure 7.17 (78698-BII)
Ontario Agricultural Museum, Milton, Ont.: Figure 3.7; Figure 3.8; Figure 3.9; Figure 3.10; Figure 3.11
Oxford University Press: Figure 1.1
R. Pequegnat: Figure 3.4; Figure 3.5; Figure 3.6
Prince Edward Island Public Archives and Record Office: Figure 5.8 (4341/1); Figure 6.6 (3466HF72.66.7.39.2); Figure 7.15 (3466/HF72.16.4.33)
Provincial Archives of New Brunswick: Figure 1.3 (P37/487/1); Figure 3.1 (P5-736); Figure 6.2 (P37-487-1); Figure 6.3 (P37-328)
Roy Archives, Peterborough: Figure 1.4; Figure 5.12
Lorne Shields: Figure 5.6
Robert Stacey, the Art Associates: Figure 1.7 (reproduced with permission)
Städtisches Reiss-Museum, Mannheim: Figure 1.2 (reproduced with permission from the full painting)
Tavistock Public Library, Ont.: Figure 1.9 (Lemp Collection); Figure 7.9 (Lemp Collection)
Toronto Public Library: Figure 1.8 (T12167); Figure 1.13 (T33495)
Vancouver Public Library: Figure 5.1 (5448); Figure 6.11 (8284)
Weldon Library, University of Western Ontario, London: Figure 1.12
Yukon Archives, Dawson City Museum: Figure 7.4 (6394, W. Harrison photo); Figure 7.5 (6521, McLennan Collection)

Index

Page references to illustrations are in italics